Katabathini Narasimharao, Huda Sharbini Kamaluddin
Heterogeneous Catalysts

Also of Interest

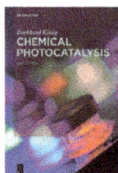

Chemical Photocatalysis
2nd Edition
Burkhard König (Ed.), 2020
ISBN 978-3-11-057654-2, e-ISBN 978-3-11-057676-4

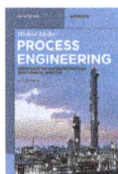

Process Engineering.
3rd Edition
Addressing the Gap between Study and Chemical Industry
Michael Kleiber, 2023
ISBN 978-3-11-102811-8, e-ISBN 978-3-11-102814-9

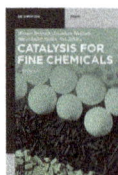

Catalysis for Fine Chemicals
Werner Bonrath, Jonathan Medlock,
Marc-André Müller, Jan Schütz, 2024
ISBN 978-3-11-109609-4, e-ISBN 978-3-11-110267-2

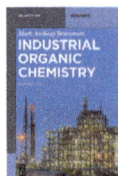

Industrial Organic Chemistry
2nd Edition
Mark Anthony Benvenuto, 2024
ISBN 978-3-11-132991-8, e-ISBN 978-3-11-133035-8

Katabathini Narasimharao,
Huda Sharbini Kamaluddin

Heterogeneous Catalysts

Metals, Oxides, Silicates, Zeolites and Nanomaterials

DE GRUYTER

Authors
Prof. Dr. Katabathini Narasimharao
Chemistry Department
Faculty of Science
King Abdulaziz University
P.O. Box 80203
Jeddah 21589
Saudi Arabia
nkatabathini@kau.edu.sa

Dr. Huda Sharbini Kamaluddin
Chemistry Department
Faculty of Science
King Abdulaziz University
P.O. Box 80203
Jeddah 21589
Saudi Arabia
hudashkamaluddin@gmail.com

ISBN 978-3-11-131679-6
e-ISBN (PDF) 978-3-11-131681-9
e-ISBN (EPUB) 978-3-11-131719-9

Library of Congress Control Number: 2024947823

Bibliographic information published by the Deutsche Nationalbibliothek
The Deutsche Nationalbibliothek lists this publication in the Deutsche Nationalbibliografie;
detailed bibliographic data are available on the internet at http://dnb.dnb.de.

© 2025 Walter de Gruyter GmbH, Berlin/Boston, Genthiner Straße 13, 10785 Berlin
Cover image: Katabathini Narasimharao and Huda Sharbini Kamaluddin and Kautsar Sarbini
Typesetting: Integra Software Services Pvt. Ltd.

www.degruyter.com
Questions about General Product Safety Regulation:
productsafety@degruyterbrill.com

Contents

Chapter 3
Different spectroscopy techniques for characterization of materials —— 77

Chapter 4
Green synthesis methods —— 143

Chapter 5
Grand challenges and opportunities in scale-up of material synthesis —— 163

Chapter 1
Introduction of solid-state materials as heterogeneous catalysts

1.1 Introduction

Jöns Jacob Berzelius, a Swedish chemist, first used the term "catalysis" in 1838, derived from the Greek *kata-* (down) and *lyein* (loosen). The catalyst process consists of two fundamental components: the catalyst material and the reaction mixture. Depending on the phase of the system, catalysis can be classified as homogeneous or heterogeneous. **Homogeneous catalysis** occurs in a single phase, such as when the catalyst and reactants are both in a solution. In contrast, **heterogeneous catalysis** involves two or more phases, such as a solid catalyst and the reactants are in a different phase (i.e., either gas or liquid) [1, 2]. Most industrial-scale catalysis is typically carried out using **heterogeneous catalysis** because it offers a straightforward separation of the catalyst, reactants, and products at the end of the reaction, making the catalyst reusable and recyclable. Furthermore, heterogeneous catalysts is generally simpler to prepare and handle [1, 3] (Figure 1.1). To understand the role of catalyst in the reaction mechanism, it is essential to obtain information about the nature and functionality of the catalyst.

	Homogenous catalysts	Heterogenous catalysts
Major feature	(i) Same phases for (catalysts, reactants, and products) (ii) Co dissolved. (iii) High selectivity	(i) Different phases for (catalysts, reactants and products). (ii) No solvent required. (iii) Poor Selectivity.
Advantages	(i) Dissolves in reaction medium; hence all catalytic sites are available for reaction.	(i) Stable. (ii) Reusable. (iii) Use as fixed beds. (iv) Easy separation.
Disadvantages	(i) Difficult separation. (ii) Reactor corrosion. (iii) Huge waste materials. (iv) Product.	(i) Nonselective to chiral catalysis. (ii) Difficult to study, and hence reaction mechanisms often need to be discovered.

Figure 1.1: The differences between homogenous and heterogeneous catalysis (reproduced from ref. [3]).

https://doi.org/10.1515/9783111316819-001

1.2 Catalysts and catalytic process

Catalysts are typically solid materials that facilitate chemical reactions by lowering the activation energy barrier between the reactants and products, allowing for the reaction to occur under milder conditions (Figure 1.2). They play a significant role in conserving energy and producing pure products instead of mixtures [4, 5]. A key characteristic of catalysts is that they are not consumed during the reaction and do not undergo chemical changes themselves. The **catalysis/catalytic process** can be classified into two categories: positive catalysis, which accelerates the reaction rate, and negative catalysis, which slows down the reaction rate [6]. The catalytic process is influenced by the activation energy (E_a), which refers to the least amount of energy needed for the molecule involved in the reaction; refer to Figure 1.2a and eq. (1.1).

$$E_a = \text{Threshold energy} - \text{Average kinetic energy of reacting molecular} \qquad (1.1)$$

Activation energy (E_a) plays a crucial role in the catalytic process. The presence of a catalyst material can affect the value of E_a. A positive catalyst decreases the activation energy of the reagent, which in turn, speeds up the reaction (Figure 1.2b). On the other hand, a negative catalyst increases the activation energy, which leads to a slower reaction rate (Figure 1.2c).

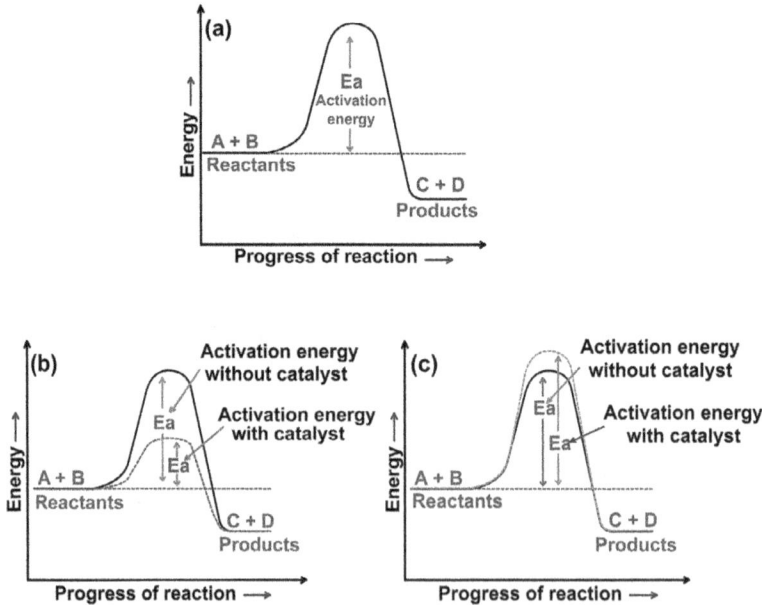

Figure 1.2: Activation energy diagram with representative catalytic system.

It is important to have a thorough understanding of the fundamental principles of solid-state materials employed in the heterogenous catalysis processes.

1.3 Structure of crystalline solids

While it is true that most solid catalysts used on a large scale are **inorganic** such as solid materials that do not contain carbon atoms bonded to hydrogen, such as minerals, metals, ceramics, and salt. Inorganic solids possess properties and structures that are dependent on their chemical composition and the arrangement of atoms. Due to the existence of molecular-scale pores and a well-defined crystalline structure, inorganic materials, which act as a transitional bridge between solutions, gels, and other solids, play a significant role in the catalysis process. The solvent-like environments present in inorganic materials serve as the location where the catalytic process occurs.

Inorganic materials that form **crystals** are structured with atoms, ions, or molecules arranged in a periodic order, represented by a three-dimensionally repeating entity known as the "**unit cell**." The solid material's structure can be categorized into single crystal, polycrystalline, or amorphous forms, as illustrated in Figure 1.3. Single crystals exhibit an atomic arrangement that indefinitely repeats unit cells, while polycrystalline solids possess grains, which can be either single crystals or multiple interconnected regions where the long-range order of atomic structure is defined. The distinct crystalline phases within a polycrystalline material are referred to as grain boundaries. For amorphous materials, the atomic arrangement is not periodic, and the distance between atoms remains fixed, resulting in short-range order. This contrasts with single crystal and polycrystalline structures, which exhibit long-range order in their atomic arrangements.

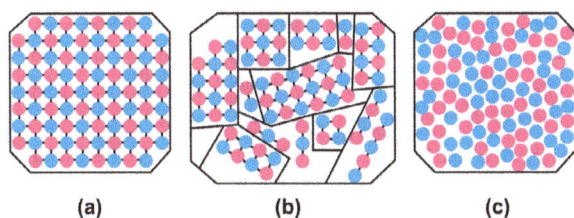

(a)   (b)   (c)

Figure 1.3: Represent different crystalline structure types: **(a)** crystalline (single crystal), **(b)** polycrystalline (grains), and **(c)** amorphous (solid-liquid).

1.3.1 Single crystal structures (crystalline crystals)

A **crystal structure** is an actual object that contains atoms, molecules, or ions arranged in a periodic, long-range order with translation symmetry. Simultaneously, a **lattice structure** represents a mathematical concept describing how points are distributed in space periodically and symmetrically. A crystal structure can be derived from a lattice structure by specifying a basis for each lattice point. A **basis** refers to a grouping of atoms, molecules, or ions located at a specific position relative to the lattice points and represents the smallest repeating unit of a crystal. This unit is called a **unit cell**. A unit cell represents the smallest repeating unit of a crystal, containing information about its crystal structure. A unit cell is defined by three vectors (a, b, and c cell dimensions) and three angles (α, β, and γ) that determine its shape and size. Moreover, it typically contains one or more **lattice points**, which are positions in space where atoms, molecules, or groups of atoms can be placed. For instance, a sodium chloride (halite) crystal structure can be constructed by attaching sodium (Na) and chlorine (Cl) atoms to each lattice point of a cubic lattice structure (Figure 1.4).

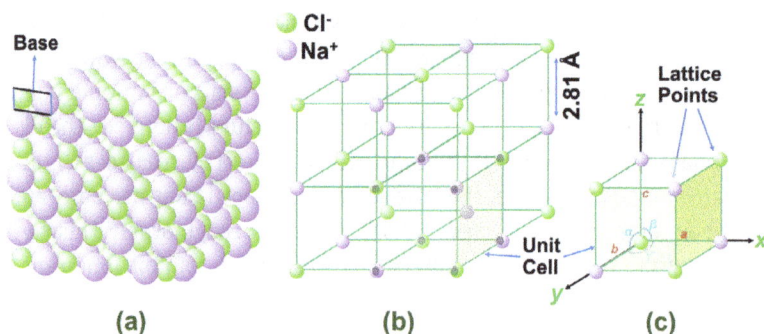

Figure 1.4: Halite sodium chloride (NaCl) crystal packing. **(a)** The set of Na and Cl atoms in the form of a crystal structure. **(b)** The Na and Cl atoms in a three-dimensional cubic lattice structure model. **(c)** A unit cell represents the locations of lattice points in the repeating system in all directions for the NaCl crystal.

Bravais lattices, named after the French physicist Auguste Bravais, describe the different arrangements observed in **single crystal structures** in nature. He was the first to describe them in 1850. **Bravais lattices** classify unit cells into seven shapes: cubic, tetragonal, orthorhombic, monoclinic, triclinic, hexagonal, and rhombohedral, creating a total of 14 distinct crystal structures (see Figure 1.5).

I.   **Cubic:** The unit cell is in a shape of a cube, and the cell dimensions are $a = b = c$, and the angles are $\alpha = \beta = \gamma = 90°$. There are three types of cubic Bravais lattices: primitive (P), body-centered (I), and face-centered (F). Examples of cubic crystals are sodium chloride, diamond, and copper.

II. **Orthorhombic:** The unit cell is in a shape of a rectangular prism with base and lattices result from stretching a cubic lattice along two different vectors. The cell dimensions are $a \neq b \neq c$, and the angles are $\alpha = \beta = \gamma = 90°$. There are four types of orthorhombic Bravais lattices: primitive (P), body-centered (I), base-centered (C), and face-centered (F). Orthorhombic crystals include rhombic sulfur, magnesium sulfate, and potassium nitrate.

III. **Tetragonal:** The unit cell is a rectangular prism shape with a square base and lattices result from stretching a cubic lattice along one of its lattice vectors. The cell dimensions are $a = b \neq c$, and the angles are $\alpha = \beta = \gamma = 90°$. There are two types of tetragonal Bravais lattices: primitive (P) and body-centered (I). Examples of tetragonal crystals are tin oxide, zircon, and rutile.

IV. **Hexagonal:** The unit cell is a hexagonal prism shape with 12-point groups. The cell dimensions are $a = b \neq c$, and the angles are $\alpha = \beta = 90°$ and $\gamma = 120°$. There is only one type of hexagonal Bravais lattice: primitive (P). Examples of hexagonal crystals are graphite, quartz, and ice.

V.   **Triclinic:** The unit cell is a parallelepiped with no symmetry and lattice points at the corners. The cell dimensions are $a \neq b \neq c$, and the angles are $\alpha \neq \beta \neq \gamma \neq 90°$. There is only one type of triclinic Bravais lattice: primitive (P). Examples of triclinic crystals are copper sulfate, feldspar, and turquoise.

VI. **Monoclinic:** The unit cell is a parallelepiped with a rectangular base and lattice points at the corners and in special cases at the center of the base or the opposite face. The cell dimensions are $a \neq b \neq c$, and the angles are $\alpha = \gamma = 90°$ and $\beta \neq 90°$. There are two types of monoclinic Bravais lattices: primitive (P) and base-centered (C). Examples of monoclinic crystals are monoclinic sulfur, sodium sulfate, and gypsum.

VII. **Rhombohedral (or trigonal):** The unit cell is a rhombohedron in shape with lattice points at the corners and in special case at the center. The cell dimensions are $a = b = c$, and the angles are $\alpha = \beta = \gamma \neq 90°$. There is only one type of rhombohedral Bravais lattice: primitive (P). Examples of rhombohedral crystals are calcite, corundum, and hematite.

The crystal structures' different building blocks are arranged in **compact packing**. Compact packing is defined as achieving the lowest possible volume for atoms in space, assuming that atoms are hard, rigid spheres. Layers are stacked in three-dimensional space (Figure 1.6). Each sphere has six nearest neighbors. The second layer can be placed over either of the "1" or "2" voids (refer to Figure 1.6) to form a compact packing arrangement.

The ideal close-packing structures for atoms or ions can be summarized in two types: **cubic close packing** (cp) and **hexagonal close packing** (hcp). Both of these arrangements are depicted in Figure 1.7. The location of the third layer offers two possibilities. The first option involves placing the spheres of the third layer directly above the centers of the first layer, forming an ABAB pattern, resulting in a 74% pack-

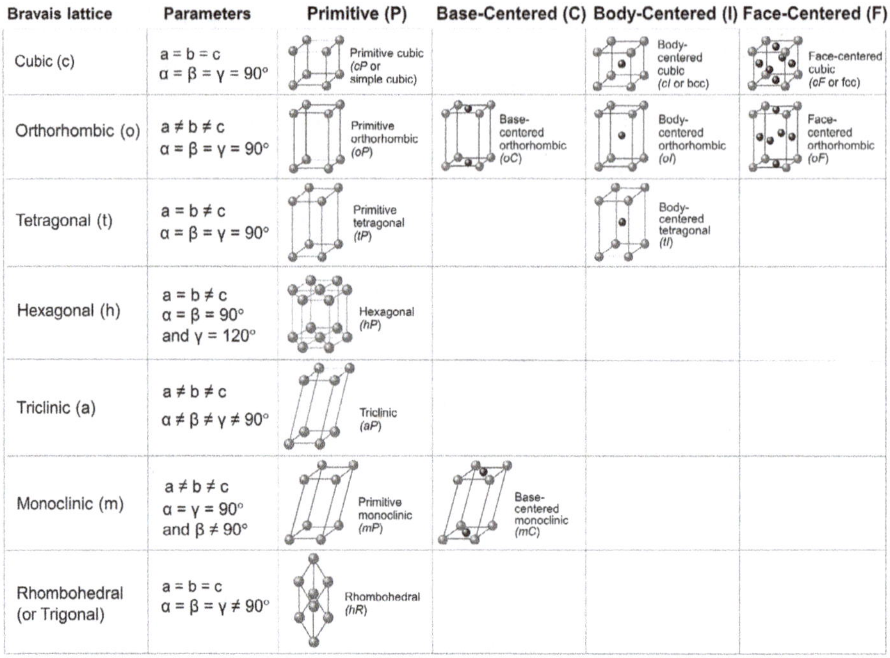

Bravais lattice	Parameters	Primitive (P)	Base-Centered (C)	Body-Centered (I)	Face-Centered (F)
Cubic (c)	$a = b = c$ $\alpha = \beta = \gamma = 90°$	Primitive cubic (cP or simple cubic)		Body-centered cubic (cI or bcc)	Face-centered cubic (cF or fcc)
Orthorhombic (o)	$a \neq b \neq c$ $\alpha = \beta = \gamma = 90°$	Primitive orthorhombic (oP)	Base-centered orthorhombic (oC)	Body-centered orthorhombic (oI)	Face-centered orthorhombic (oF)
Tetragonal (t)	$a = b \neq c$ $\alpha = \beta = \gamma = 90°$	Primitive tetragonal (tP)		Body-centered tetragonal (tI)	
Hexagonal (h)	$a = b \neq c$ $\alpha = \beta = 90°$ and $\gamma = 120°$	Hexagonal (hP)			
Triclinic (a)	$a \neq b \neq c$ $\alpha \neq \beta \neq \gamma \neq 90°$	Triclinic (aP)			
Monoclinic (m)	$a \neq b \neq c$ $\alpha = \gamma = 90°$ and $\beta \neq 90°$	Primitive monoclinic (mP)	Base-centered monoclinic (mC)		
Rhombohedral (or Trigonal)	$a = b = c$ $\alpha = \beta = \gamma \neq 90°$	Rhombohedral (hR)			

Figure 1.5: The 14 Bravais lattices. These 14-unit cells generate all the possible three-dimensional crystal lattices.

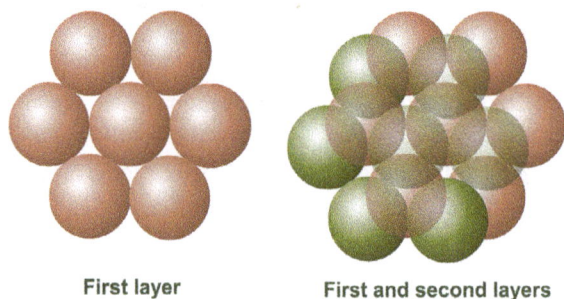

First layer   First and second layers

Figure 1.6: Close packing of spheres.

ing. The second possibility places the spheres of the third layer in the cusps of the second layer, slightly offset from the ABAB pattern, creating an ABCABC arrangement with a 74% packing. The ABAB structure, also known as hexagonal close packing, offers a hexagonal arrangement of atoms in a close-packed manner. Conversely, the ABCABC structure corresponds to cubic close packing with a face-centered cubic geometry, where atoms are arranged in a tightly packed arrangement around the center of each cubic unit.

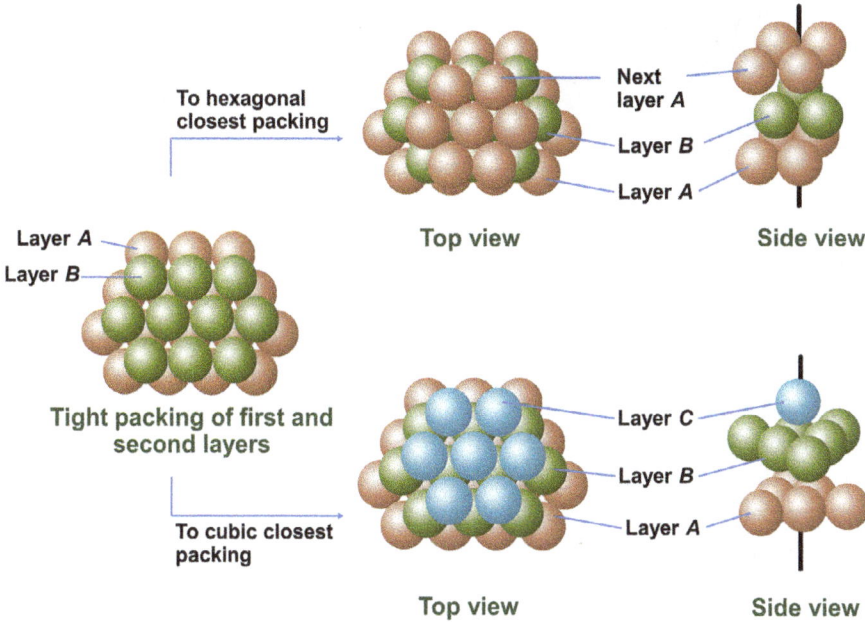

Figure 1.7: The two possibilities of close-packing spheres: hexagonal closest packing and cubic closest packing.

1.3.2 The orientation of a lattice plane by Miller indices

Miller indices represent a method to identify the orientation of a plane or surface within a crystal. They are derived from the plane's intersection with the crystal's three primary axes. These Miller indices (hkl), where h, k, and l are integers that specify the inverse ratio of the plane's intercepts with the x, y, and z axes as integers. For a cubic structure with a lattice constant "a" and spacing "d" between adjacent (hkl) lattice planes, the formula for calculating "d" is given as:

$$d_{hkl} = \frac{a}{\sqrt{h^2 + k^2 + l^2}} \tag{1.2}$$

For instance, a plane intersecting the x-axis at $a/2$ and being parallel to the y- and z-axes would be represented by the indices (200), as illustrated in Figure 1.8. The Miller indices would identify a plane intersecting the x-axis and y-axis at "a" and the z-axis at $a/2$ (112). A plane parallel to all three axes is not valid and would be represented as (000). Moreover, negative numbers are denoted with a bar over them in Miller indices, with $(12\bar{3})$ symbolizing a plane intersecting the x-axis at "a", the y-axis at $a/2$, and the negative z-axis at $\bar{a}/3$. **Miller indices notations** can be further defined as *(hkl)* for a plane, *{hkl}* for the family or set of planes, *[hkl]* for a direction, and *<hkl>* for a family of directions.

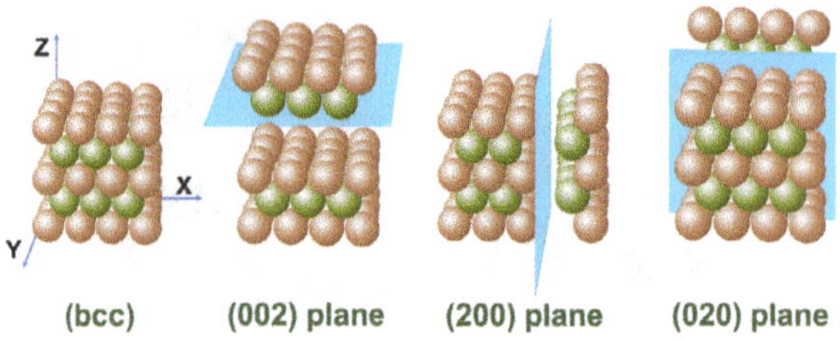

(bcc)   (002) plane   (200) plane   (020) plane

Figure 1.8: The low Miller index planes of a cubic, close-packed (body-cantered cubic) metal.

Cutting the lattice along the direction with the highest atomic density (as determined by the lowest Miller indices) results in a more uniform crystal surface, as seen in the flat faces including (002), (200), and (020) (Figure 1.8). When the lattice is cut along the plane with the lowest atomic density (indicating higher Miller indices), a rough, heterogenous mixture of atoms with varying degrees of surface coordination and different types of unsaturation is formed, including adatoms (adsorbed atoms), terrace (face), steps (edges), and kink (corners) structures (Figure 1.9) [7]. Changes in the cleavage angles induce changes in the width of terraces and the density of steps. These different types of unsaturated atoms or molecules, also known as interfacial species (interference), play a crucial role in catalytic processes. Variations in cleavage angles give rise to terrace width and step density changes. Terraces, steps, and kinks have inherent stability, though surface reconstruction can occasionally occur under a vacuum, affecting only the top few layers. Compared to simplistic structures, surfaces with steps and kinks resemble typical catalyst surfaces, replete with nonuniformities and imperfections.

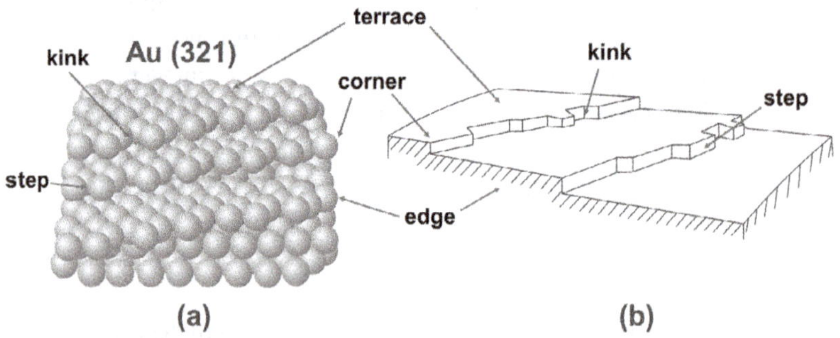

(a)   (b)

Figure 1.9: (a) Surface chirality of high Miller index planes of Gold (Au) single crystal with different terrace, step, and kink. Adapted from ref. [8] **(b)** A microscopic view of the Au metal surface shows several site geometries [9].

Single-crystal surfaces serve as excellent models for regular surfaces, and the step and kink structures demonstrate the nonuniformities (heterogeneity) at an atomic scale. Other forms of nonuniformities (or defects) exist on surfaces, such as point defects resulting from adsorbed atoms or vacancies in the lattice. Large-scale surface defects are common, with some being detectable through microscopy. Dislocations in single crystal surfaces arise from mismatches between atomic planes, manifesting as line defects. Typical dislocation densities for metal or ionic single-crystal surfaces range from 10^6 to 10^8 per square centimeter, while densities for metal oxides often range from about 10^4 to 10^6 per square centimeter.

1.3.3 Surface and interface in solid catalyst

Chemical reactions occur at the solid surface and interface regions in heterogeneous catalyst systems. Depending on the reaction phase, these interfaces encompass three interfaces involving solid-gas, solid-liquid, and solid-gas-liquid combinations. Catalysts provide active sites where reactants are adsorbed, activated, and transformed into new chemical species before being released from the catalyst surface. The catalyst's capacity to facilitate these reactions is contingent upon the surface thermodynamic binding energy resulting from the coordination of unsaturated atoms on the surface [10], which can be modified by tailoring the atomic structure of the interfacial atoms or by creating new interfaces.

The majority of solid catalysts are polycrystalline (i.e., grainy) (Figure 1.3b) and display distinct facets with varying surface atomic structures, edges, corners, and defects, leading to many active sites. Consequently, a catalyst particle is expected to expose multiple types of active sites, with its catalytic performance represented as the aggregate of the individual catalytic performances of each active site type. The task of identifying the structures and densities of active sites presents significant challenges in *in situ* studies due to the unavailability of surface characterization techniques that can function under catalytic reaction conditions [10]; it becomes essential to design solid catalysts from the outset meticulously. For instance, acid removal alters the chemical distribution of atomic arrangements and catalyst surfaces following synthesis. Incorporating core-shell structures has demonstrated utility in adjusting surface tensions and surface electronic structures. Structural twins and internal phase boundaries within single catalyst particles have been found to modify the local surface atomic configuration, leading to variations in catalyst activity. Grasping the function of these interfaces in catalytic reactions, particularly how their atomic and chemical configurations influence reaction mechanisms and how they evolve under dynamic electrochemical processes, is pivotal in devising new catalysts with optimal performance.

1.3.4 Criteria for choosing industrial catalyst

Usually, industrial catalysts require a delicate blend of distinct properties during both the synthetic process and the catalytic process, and they should be chosen according to the following criteria: (i) the catalysts materials should be easily synthesized, (ii) should be cost-effective, (iii) a bounder widely, and (iv) reproducible. During catalysis, the following characteristics should be present in the catalyst material: (i) high catalytic activity, (ii) high selectivity, (iii) proper pore structure, (iv) an extended lifetime, (v) high resistance to deactivation and catalytic poisons, (vi) easy regeneration, (vii) low operating temperature to facilitate catalyst activation/deactivation, (viii) high thermal stability, (ix) thermal conductivity, (x) mechanical strength, and (xi) resistance to attrition [11].

1.3.4.1 Synthesize of catalyst materials
The objective of a catalyst manufacturer is to generate a commercially viable product that functions as a stable, highly active, and selective catalyst throughout the catalytic process. In order to accomplish this, industrial catalysts should optimally be synthesized under easy conditions, such as ambient temperature and pressure, utilizing priceless reagents abounding naturally. However, the more intricate production procedure could be more favorable for commercial production and reproducibility when taking into account the precious resources of time, budget, and trial-and-error. Consequently, industrial catalyst materials are most commonly generated using these two common methods: (i) blending and (ii) mounting. In the blending approach, they typically employ conventional precipitation techniques (with the alternative of utilizing co-precipitation). This process leads to the creation of a new solid phase using appropriate reagents (known as precipitating agents) from a liquid medium; the resulting precipitate is subsequently transformed into the active catalyst during subsequent preparation phases. Conversely, the mounting method involves the impregnation route, where a solid phase is first performed separately and functions as a support, and the catalytically active material is then mounted and stabilized on it. In this approach, the part of the catalyst material possesses mechanical properties influenced by the preexisting support, and the preparation process centers on introducing the catalytic compounds. From a structural perspective, the simplest types of catalysts widely used in the industry include (i) bulk metals and alloys, (ii) bulk oxides, (iii) sulfides, (iv) carbides, (v) borides, and (vi) nitrides. This is because those materials display uniformity at the molecular level, exposing catalytic properties on the external surface and facilitating material development in diverse environments [12].

1.3.4.2 Catalytic processes
The appropriate catalyst material must catalyze desirable products at high rates and minimal costs with high activity, selectivity, and stability. The activity required to

achieve high conversion under mild operating conditions, such as lower temperature and pressure in a small reactor volume. The catalyst's selectivity represents its ability to produce the desired product over alternative products and is typically expressed as a percentage of total product output. Additionally, catalyst stability denotes how long the activity and selectivity will be sustained under specific operating conditions. Catalyst stability affects the cycle length, which is defined based on the severity of the operation. Consequently, various feedstock scenarios lead to different cycle lengths. It is crucial to comprehend the factors impacting catalyst lifetime to make an informed decision on managing challenging feedstock molecules and accepting a shorter cycle time or omitting them from the feedstock to achieve a longer cycle length.

1.4 References

[1]   G. Palmisano, S. Al Jitan and C. Garlisi. *Heterogeneous Catalysis: Fundamentals, Engineering and Characterizations*, Elsevier, 2022.

[2]   P. Unnikrishnan and D. Srinivas, Heterogeneous catalysis: in *Industrial Catalytic Processes for Fine and Specialty Chemicals*, eds. S. S. Joshi, V. V. B. T.-i. C. P. For F. and S. C. Ranade, Elsevier, Amsterdam, 2016, pp. 41–111.

[3]   M. E. Ali, M. M. Rahman, S. M. Sarkar and S. B. A. Hamid. Heterogeneous metal catalysts for oxidation reactions, *J. Nanomater*, 2014, **2014**, 192038.

[4]   S. Ashworth. Catalytic metals and their uses, https://edu.rsc.org/feature/catalytic-metals-and-their-uses/3007558.article, (accessed 7 July 2023).

[5]   R. Schlögl. Heterogeneous Catalysis, *Angew. Chemie Int. Ed*, 2015, **54**, 3465–3520.

[6]   P. Sahu. Catalysis: Definition, types of catalyst and example, https://www.embibe.com/exams/catalysis/, (accessed 12 July 2023).

[7]   R. L. Augustine. Whither goest thou, catalysis, *Catal. Letters*, 2016, **146**, 2393–2416.

[8]   S. W. Im, H.-Y. Ahn, R. M. Kim, N. H. Cho, H. Kim, Y.-C. Lim, H.-E. Lee and K. T. Nam. Chiral surface and geometry of metal nanocrystals, *Adv. Mater*, 2020, **32**, 1905758.

[9]   J. Ryczkowski. Scientific co-operation with professor Nazimek, *Sect. AA*, 2010, **16**, 94–105.

[10]   W. Huang and W.-X. Li. Surface and interface design for heterogeneous catalysis, *Phys. Chem. Chem. Phys*, 2019, **21**, 523–536.

[11]   L. Petrov, Problems and challenges about accelerated testing of the catalytic activity of catalysts: in *Principles and Methods for Accelerated Catalyst Design and Testing*, eds. E. G. Derouane, V. Parmon, F. Lemos and F. R. Ribeiro, Springer, Netherlands, Dordrecht, 2002, pp. 13–69.

[12]   J. A. Schwarz, C. Contescu and A. Contescu. Methods for preparation of catalytic materials, *Chem. Rev*, 1995, **95**, 477–510.

Chapter 2
Surface chemistry and different synthesis strategies

2.1 Surface chemistry

The heterogeneous catalyst process occurs on the surface of a solid material system. Depending on the phase of the reactant, the *reactive interfaces* can include those between solid-gas, solid-liquid, and triple interfaces of solid-gas-liquid. At this interface, the catalyst provides active sites where one or all reactant species must come near the surface forming bond or, at a minimum, interact with it, called the adsorption process. These phenomena help to understand the mechanism of the catalytic process. Furthermore, the catalyst catalyzes the conversion of reactants into new chemical species that are subsequently released from the catalyst surface. Based on Langmuir's pioneering work, he observed adsorption phenomena through the coincidence of light bulb usage. He noted that when hydrogen (H_2) comes into contact with a hot Tungsten (W) filament, it dissociates into hydrogen atoms and forms a one-layer thick on the surface of the light bulb.

2.2 Adsorption process steps

As we know, the heterogeneous catalytic reaction occurs on the surface of a solid catalyst through adsorption. The term "adsorbed" indicates uptake, where the reactant (referred to as the adsorptive) approaches (adsorbent) and interacts with the catalyst's surface (referred to as the adsorbed). After the reaction, the newly formed product or products must detach from the catalyst surface and move away from it in a process term as desorption. The process is illustrated in Figure 2.1a. The process of a heterogeneous catalytic reaction can be described in five steps, which is depicted in Figure 2.1b.

1. Reactant diffusion to active surface,
2. Reactant adsorption onto the surface,
3. Reaction on the surface,
4. Product desorption from the surface, and
5. Product diffusion away from the surface [1].

https://doi.org/10.1515/9783111316819-002

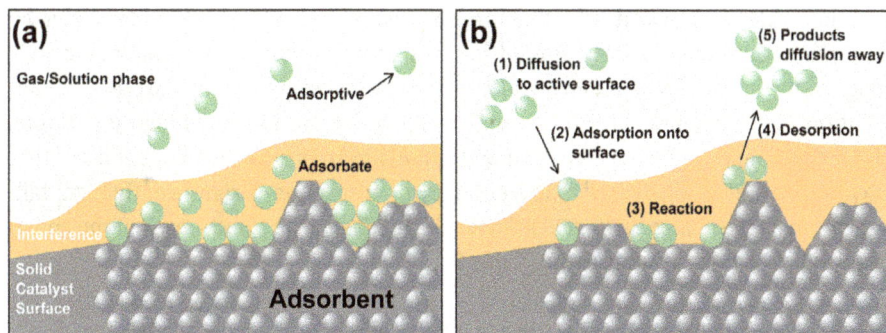

Figure 2.1: (a) Adherence of molecules (or ions and atoms) to the active surface of a solid is known as adsorption. **(b)** Schematic illustration of the adsorption process.

2.3 Physical and chemical adsorption

Adsorption is the process of condensing or interacting and forming bonds between a specific species (referred to as adsorbents) on the surface of a solid material (referred to as the adsorbent). It is classified into two primary types: physical adsorption (or physisorption) and chemical adsorption (or chemisorption). This differentiation is based on how the chemical species interacts with the adsorbent surface. In all adsorption processes, thermal energy is released to the surroundings, resulting in a negative enthalpy ($-\Delta H$). As the gas is converted into a more restricted, and perhaps more ordered, adsorbed layer, its entropy decreases. For spontaneous adsorption, energy must be supplied to the thermal surroundings [2, 3].

In the case of physisorption, the reactant species adsorb to the adsorbent surface through weak intermolecular forces, encompassing Van der Waals forces like London dispersion and dipole-dipole interactions as well as hydrogen bonding. Consequently, the adsorbate's and adsorbent's electron density distribution remains unaltered through physisorption. The enthalpy of adsorption is generally low in physisorption, falling within a range of 5–40 kJ/mol, contingent on the molecular masses and properties of the species involved. Furthermore, physisorption is often characterized as a speedy, reversible, non-dissociative process with potential adsorbate multilayer formation. During the formation of the initial layer, the heat of adsorption is roughly parallel to, yet not rigorously equivalent to, the heat of condensation. In contrast, the heat of adsorption reflects the heat of liquefaction necessary for forming the second and subsequent layers. Physisorption typically transpires within temperature ranges close to or below the condensation point of the adsorbate. If physisorption were to occur at a temperature above the condensation point, it would be restricted to a single layer because condensing a second layer onto the first one would become impossible.

The term "chemisorption" refers to the adsorption of a species onto a surface through the breaking or formation of ionic or covalent bonds, resulting in a significant change in the electron density arrangement. Compared to physisorption, chemisorption often demands activation, making its adsorption kinetics highly variable. Chemisorption, usually limited to forming a monolayer, is an irreversible process that may or may not be dissociative. The enthalpy of adsorption is significantly high, falling within a wide range of 40–800 kJ/mol, which depends on the strength of the chemical bonds involved, as indicated in Table 2.1.

Table 2.1: Characteristics of physical adsorption and chemisorption [3].

Physisorption	Chemisorption
– Enthalpy of adsorption less than about 40 kJ/mol.	– Enthalpy of adsorption greater than about 80 kJ/mol.
– Adsorption is appreciable only at temperatures below the boiling point of the adsorbate.	– Adsorption can occur at high temperatures.
– Incremental increase in the amount adsorbed.	– Incremental increase in amount adsorbed.
– Increase with each incremental.	– Decrease with each incremental.
– Increase in pressure of adsorbate.	– Increase in the pressure of adsorbate.
– The amount of adsorption on the surface is a function that is more of an adsorbate than an adsorbent.	– The amount of adsorption characteristic on both adsorbate and adsorbent.
– No appreciable activation energy is involved in the adsorption process.	– Activation energy may be involved in the adsorption process.
– Multilayer adsorption occurs.	– Adsorption leads to a monolayer, at most.

The outermost layer of an adsorbate may undergo rearrangement upon the formation of an adsorbate monolayer. The structure of the adsorbent system influences the bond strength of adsorbates. Single-crystal metal particles, particularly those with high-index planes, are recognized to have a high density of atomic-sized steps that serve as potent catalytic sites. The bond strength may change due to alterations in the structure at the reaction site (atomic structure effect) or modifications in the structure elsewhere in the system (electronic structure effect). These effects can lower the energy barrier for dissociating a diatomic molecule at a stepped metallic surface.

2.4 Types of adsorption process

It is an equilibrium distribution between the adsorbent's surface and the adsorbate's surrounding gas phase. Five factors can influence this equilibrium between the adsorbent and the adsorbate, namely (i) pressure, (ii) temperature, (iii) the surface area of the adsorbent, (iv) the nature of the adsorbent, and (v) the nature of the adsorbate.

When the amount of gas adsorbed on the surface of the adsorbent is considered at a constant temperature and is expressed as a function of relative pressure, this is known as an adsorption isotherm. In contrast, if the amount of gas adsorbed on the surface of the adsorbent is examined at constant pressure, the function of temperature is referred to as an adsorption isobar. When the amount of adsorbate adsorbed on the surface of the adsorbent remains constant, and both the pressure and temperature are functional, this is called an adsorption isostere (see Figure 2.2).

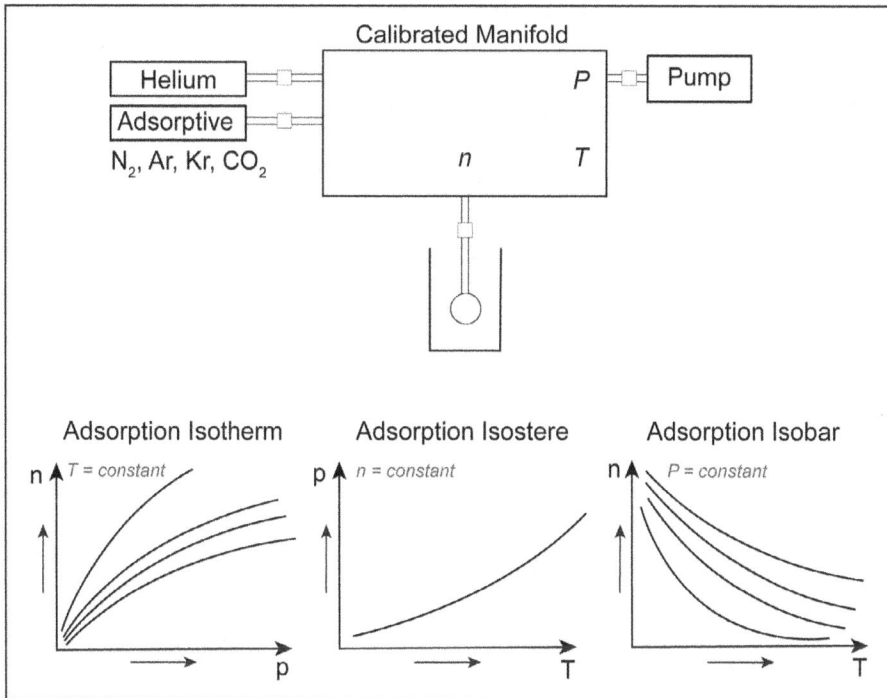

Figure 2.2: A schematic illustration of different types of adsorption processes, highlighting their dependence on temperature (T), pressure (P), and amount of adsorbent (n).

2.4.1 Adsorption isostere

The plot (graph) illustrates the relationship between pressure and temperature at a constant value of the amount, or excess amount, of substance adsorbed by a given amount of solid or a constant amount of adsorbed gas. In this experimental procedure, the heat of adsorption, $-\Delta H$, was used for assessment. This was chosen over calorimetry because direct detection of heat effects in non-ambient conditions is complicated and more expensive [4]. Once adsorption/desorption equilibrium had been established, the heat of adsorption, $-\Delta H$, for a coverage θ, was determined by analyzing adsorption isotherms measured at different temperatures. The Clausius-Clapeyron equation was then applied:

$$\frac{d(\ln P)}{d(1/T)} = \frac{-\Delta H}{R} \tag{2.1}$$

A plot of **ln P** versus **1/T** can be created, with the slope providing the isosteric heat of adsorption at a constant coverage (Figure 2.3). The magnitude of the heat of adsorption can provide insights into the nature of the adsorbate-adsorbent linkage. For instance, a heat of adsorption that surpasses the heat of liquefaction may imply a chemisorption process, where the adsorbate becomes chemically bonded to the surface of the adsorbent [1].

For example, Men'shchikova et al. studied the distribution of micropores using isosteres of methane adsorption in activated carbon (AC). Linear adsorption isosteres in the region where gases significantly deviate from ideality, see Figure 2.3a, indicating a specific state of a highly dispersed substance in micropores. This state allows methane to accumulate in micropores without undergoing phase transition over broad intervals of sub- and supercritical temperatures and pressures [5]. Chang Tao et al. studied the adsorption isosteres of methane on shale samples and found that the average isosteric heat of adsorption was approximately 20.93 kJ/mol (Figure 2.3b). This value is significantly lower than the heat required for chemical adsorption, suggesting that the adsorption process, in this case, might be primarily physical. The isosteric heat appeared to increase proportionally to the volume of adsorbed gas [6].

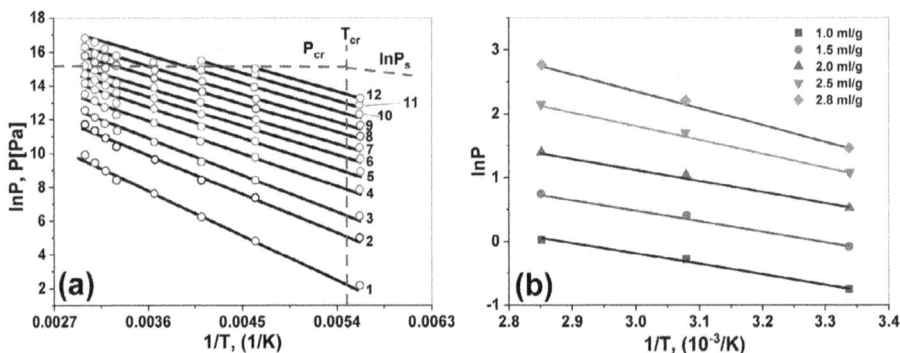

Figure 2.3: (a) The isosters of methane (CH_4) adsorption in AC at various of methane adsorption (mmol/g: **(1)** 0.1, **(2)** 0.5, **(3)** 1.0, **(4)** 2.0, **(5)** 3.0, **(6)** 4.0, **(7)** 5.0, **(8)** 6.0, **(9)** 7.0, **(10)** 8.0, **(11)** 9.0, and **(12)** 9.5. The bold dashed line shows ln P_s, where P_s is the saturated vapor pressure; the dashed lines show the critical pressure and temperature of methane (reproduced from ref. [5]). **(b)** Adsorption isosteres of methane (CH_4) on shale samples (reproduced from ref. [6]).

2.4.2 Adsorption isobar

Functional studies relate to the amount, mass, or volume of substance adsorbed by the adsorbent and the temperature at a constant pressure. These studies typically fall into two categories: physical adsorption and chemical adsorption.

For physisorption, which typically occurs at temperatures below 26.85 °C, an increase in temperature generally leads to a decrease in the amount of adsorption, as shown in Figure 2.4a. This phenomenon can be attributed to the increased kinetic energy of adsorbate molecules, which reduces the probability of their adhesion upon collision with the adsorbent surface. Consequently, a decrease in the amount of adsorption is anticipated. It is also worth noting that adsorption is an exothermic process, which implies that it releases heat as the adsorption occurs. Thermodynamically aspect, this process becomes less favorable at higher temperatures.

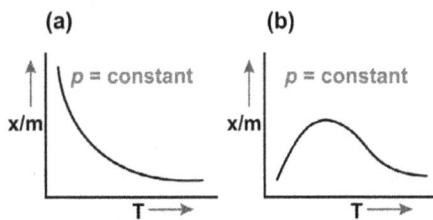

Figure 2.4: The adsorption isobar for adsorption at constant pressure: **(a)** physisorption and **(b)** chemisorption.

In contrast, chemisorption typically occurs at temperatures greater than 26.85 °C. Initially, an increase in temperature increases the amount of adsorption, as illustrated in Figure 2.4b. This is due to kinetic constraints associated with low temperatures and the necessity of overcoming the activation energy required for chemisorption, an activated process involving the breakage and formation of new bonds, a chemical reaction. However, following the activation barrier being exceeded, a subsequent increase in temperature leads to a decrease in the adsorbed amount. The isobar depicted in Figure 2.5a demonstrates the shift from physisorption to chemisorption that occurs following overcoming the chemisorption energy barrier.

Let's consider a molecule adsorbing dissociative onto a surface and breaking into two parts. In this scenario, and at low temperatures, the surface would adsorb twice as many molecular species as dissociated ones. This is because the system minimizes energy, and molecular adsorption is favored. However, as temperature increases, entropy effects become more significant and non-negligible. When the temperature exceeds 26.85 °C, entropy effects are substantial enough to surpass the energy difference between molecular and dissociative adsorption. Consequently, the adsorption process shifts from molecular to dissociative, as depicted in Figure 2.5b.

(a) **(b)**

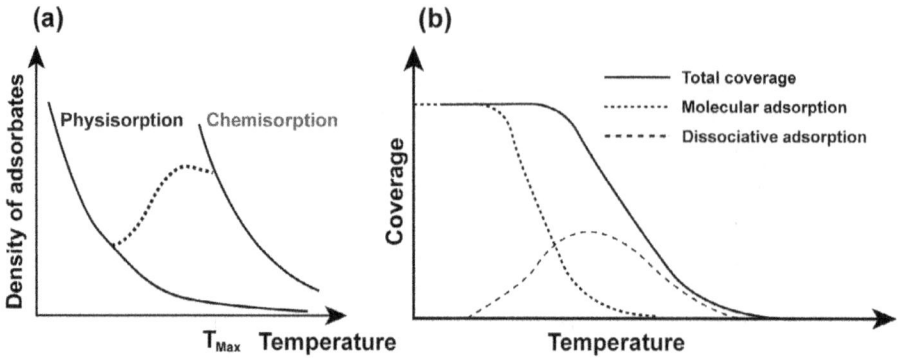

Figure 2.5: (a) A typical adsorption isobar demonstrates the shift from physisorption to chemisorption after overcoming the chemisorption energy barrier. Solid lines represent equilibrium isobars, while the dashed line represents irreversible chemisorption. Chemisorption T_{max} represents the temperature at which the maximum chemisorption coverage may be achieved (reproduced from ref. [7]) and **(b)** adsorption isobars for dissociative and molecular adsorption (reproduced from ref. [1]).

For instance, Rahman et al. [8] evaluated the thermodynamic formulation of adsorbed phase-specific heat capacity ($c_{p,a}$) derived from equilibrium uptakes of methane adsorption onto AC. Figure 2.6 illustrates the plotted adsorbed phase-specific heat capacity ($c_{p,a}$) values against varying temperatures at different constant pressures.

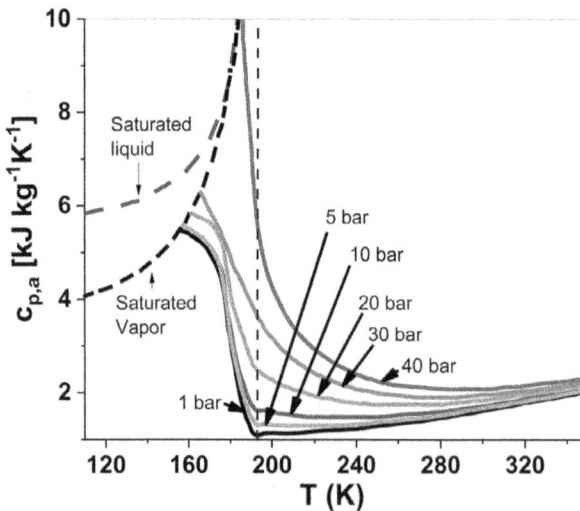

Figure 2.6: Isobars for the adsorbed phase-specific heat capacity ($c_{p,a}$) of methane adsorption onto activated carbon (reproduced from ref. [8]).

2.4.3 Adsorption isotherm

The most common adsorption experiment involves measuring the amount of adsorbed gas relative to the gas pressure at a constant temperature, and the results are typically presented graphically as adsorption isotherms. Experimentally, this is done by measuring the volume of gas adsorbed by a given amount of adsorbent or monitoring the change in the weight of the adsorbent when exposed to gas at a given pressure. Several devices are available for performing adsorption experiments: (i) gravimetric and (ii) volumetric, as shown in Figure 2.7. Adsorption isotherms can generally be described by theoretical equations using various models including (1) Langmuir, (2) Henry, (3) Freundlish, (4) Temkin, (5) Brunauer-Emmett-Teller (BET), and (6) Polanyi.

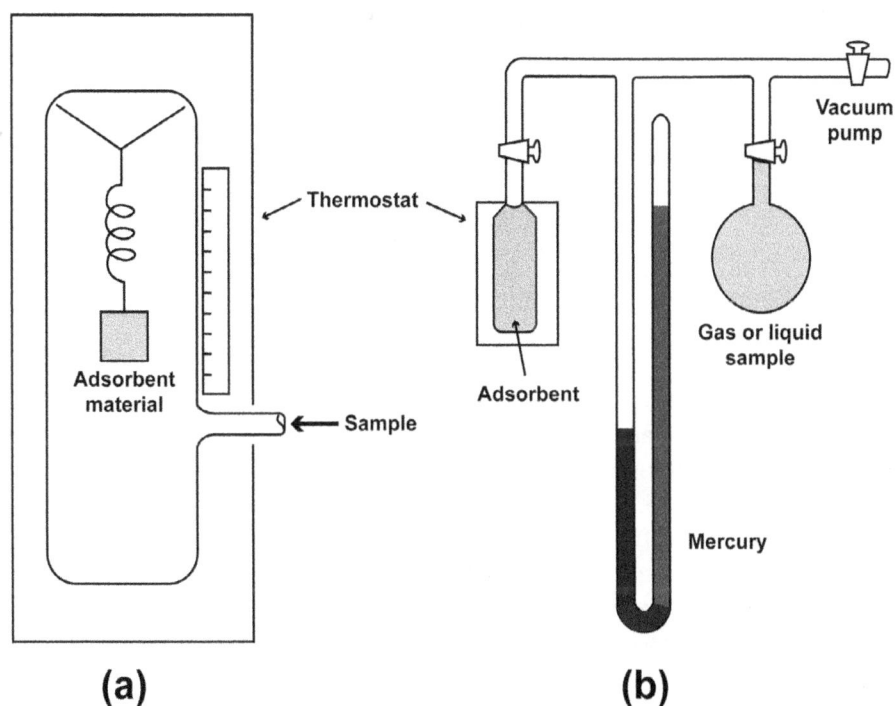

(a)   **(b)**

Figure 2.7: Devices for measuring adsorption isotherms: (a) gravimetric; (b) volumetric.

2.4.3.1 The Langmuir isotherm

In 1916, Langmuir introduced a model for the adsorption process, particularly for the chemisorption process, and derived a simple yet significant theoretical adsorption isotherm. The chemisorption process is depicted as ultimately leading to a monomolecular film formation over the surface of the adsorbent. The derived adsorption isotherm results from the equilibrium between the gas phase and the partially formed monolayer, where the fractional covering surface is represented by θ at a given gas pressure (P). The equilibrium state can be interrupted in terms of the dynamic equilibrium that results from an equal rate of desorption of the adsorbed species and the rate of condensation of gas-phase molecules.

The Langmuir theory proposes that the desorption rate is proportional to the surface fraction covered and can be written as $k_1\theta$, where k_1 is a proportionality constant. (This simple proportionality assumption neglects the fact that the enthalpy of adsorption typically depends on the extent of coverage.) According to the kinetic-molecular theory, the gas pressure P determines the frequency of molecular collisions per unit area, while the fraction of the surface not already covered by adsorbed molecules is represented by $1 - \theta$. Only collisions with the exposed surface are assumed to result in molecules sticking to the surface. The relationship between equilibrium surface coverage and gas pressure is then determined by equating expressions derived for the rate of evaporation and the rate of condensation, that is

$$k_1\theta = k_2 P \ (1-\theta) \tag{2.2}$$

where k_2 is another proportionally constant. Rearrangement gives

$$\theta = \frac{k_2\ P}{k_1 + k_2\ P} \tag{2.3}$$

Introduction of $\alpha = k_1/k_2$

$$\theta = \frac{P}{\alpha + P} \tag{2.4}$$

Inspection of eqs. (2.2) and (2.3) reveals that this theory yields a chemisorption-type isotherm at low gas pressures, where P in the denominator can be disregarded compared to α. In such cases, eqs. (2.2) and (2.3) simplify to a direct proportionality with the increase in P. For sufficiently high gas pressures, the equilibrium surface coverage (θ) approaches a constant value of unity. Experimental isotherm data encompasses information on the quantity of gas adsorbed by a specified amount of adsorbent as a function of the gas pressure. For adsorption up to the formation of a monolayer, the amount of gas adsorbed (y) at a given pressure (P), and the quantity of gas necessary to form a monolayer (y_m) can be related to surface coverage (θ) through the following equation:

$$\frac{y}{y_m} = \theta \tag{2.5}$$

Equation (2.4) becomes

$$\theta = \frac{y_m P}{\alpha + P} \tag{2.6}$$

Experimental results can be evaluated in comparison with Langmuir's theory more effectively if eq. (2.6) is rearranged to the following form:

$$\frac{P}{y} = \frac{a}{y_m} + \frac{P}{y_m}$$

(2.7)

If the experimental data align with the Langmuir theory, a plot between P and P/y should produce a straight line. If this is the case, the intercept can be identified with a/y_m and the slope with the constant $1/y_m$. As shown in Figure 2.8, many chemisorptions have an excellent linear relationship when following the Langmuir-suggested plot.

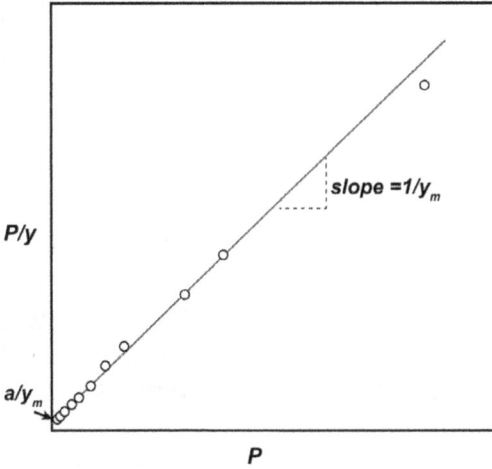

Figure 2.8: The Langmuir plot is based on the adsorption isotherm.

2.4.3.2 Competitive adsorption isotherms

When two substances (A and B) compete to adsorb on the same adsorbent surface, θ_j represents the fraction of the surface covered by molecules of species j. Consequently, $(1 - \Sigma\theta_j)$ denotes the bare fraction. The bare fraction becomes $(1 - \theta_A - \theta_B)$ in this scenario. When both species A and B are adsorbed on the surface, the expressions for the rates of adsorption and desorption can be written as follows:

$$r_a^A = k_a^A C_A \ (1 - \theta_A - \theta_B)$$

(2.8)

$$r_a^B = k_a^B C_B \ (1 - \theta_A - \theta_B)$$

(2.9)

$$r_d^A = k_d^A . \theta_A$$

(2.10)

$$r_d^B = k_d^B . \theta_B$$

(2.11)

At equilibrium, $r_a^A = r_d^A$ and $r_a^B = r_d^B$. From which we obtain

$$\frac{\theta_A}{1 - \theta_A - \theta_B} = K_A \, C_A \tag{2.12}$$

$$\frac{\theta_B}{1 - \theta_A - \theta_B} = K_B \, C_B \tag{2.13}$$

where $K_j = (k_a^j / k_d^j)$ is the adsorption coefficient for species j. Solving eqs. (2.12) and (2.13) simultaneously, we obtain

$$\theta_A = \frac{K_A \, C_A}{1 + K_A \, C_A + K_B \, C_B} \tag{2.14}$$

$$\theta_B = \frac{K_B \, C_B}{1 + K_A \, C_A + K_B \, C_B} \tag{2.15}$$

In terms of partial pressures, these two equations become

$$\theta_A = \frac{K_A \, P_A}{1 + K_A \, P_A + K_B \, P_B} \tag{2.16}$$

$$\theta_B = \frac{K_B \, P_B}{1 + K_A \, P_A + K_B \, P_B} \tag{2.17}$$

It is helpful to consider what might occur in the equations when $P_B = 0$ or $K_B = 0$. A critical assumption made in the above analysis is that both species A and B are adsorbed without undergoing molecular dissociation on the adsorbent surface [9].

2.4.3.3 Classification of adsorption isotherms

Two primary branches are observed on physisorption and chemisorption isotherms during the adsorption process when the partial pressure is gradually increased from 0 to 1 and then reversed back from 1 to 0. Several different isotherm shapes exist, with the International Union of Pure and Applied Chemistry (IUPAC) classifying the isotherms into six distinct categories, as shown in Figure 2.9, which depicts Type I to Type VI isotherms.

The physicochemical properties of the adsorbent and the porous structure influence the shape of the adsorption isotherm. Pores are classified into three categories based on their size: micropores, mesopores, and macropores. Micropores have dimensions less than 2 nm, mesopores range from 2 to 50 nm, and macropores have dimensions exceeding 50 nm [10, 11].

Type I adsorption isotherms are associated with forming a monolayer of gas molecules adsorbed on the surfaces of microporous solids at low relative pressure values (typically below 0.3). Gas adsorption in this region occurs between gas adsorbates and the walls of micropores within the adsorbent. Free open binding sites on the adsorbent surface are occupied as pressure increases, diminishing the likelihood of

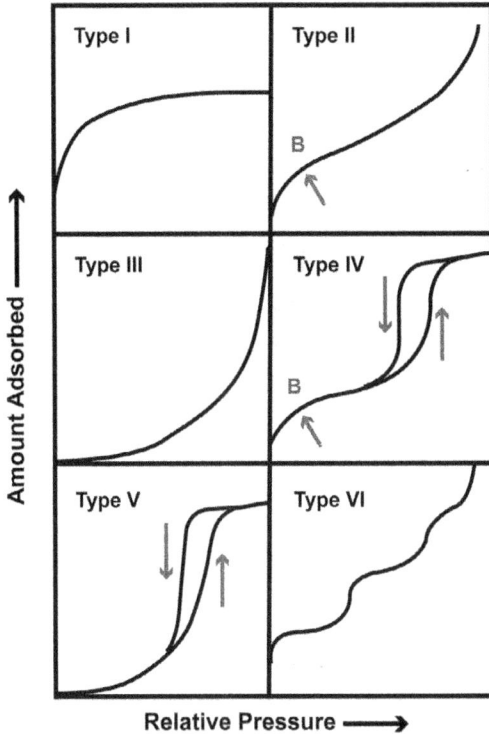

Figure 2.9: Several types of adsorption isotherms are identified within the classification system provided by IUPAC, ranging from Type I to Type VI (reproduced from ref. [1]).

further adsorption. Increasing pressure progressively fills micropores, with adsorption then proceeding on the external surface until saturation is reached, at which point no new adsorption occurs upon further pressure increases. A notable example of a microporous solid that displays this behavior is zeolites, active carbons, and zeolite-like crystalline materials. This particular adsorption isotherm shape is also characteristic of chemisorption, as demonstrated by phenomena such as oxygen adsorption on charcoal at −183 °C [12] and adsorption of the carbon dioxide (CO_2) on zeolite 13X under varying isothermal conditions [13] (see Figure 2.10a).

Type II isotherms are characterized by the adsorption of adsorbates on nonporous or macroporous solid materials. In the low-pressure region, adsorbates adsorb on open sites, forming a monolayer, with monolayer coverage completing at point B. As the pressure increases, the likelihood of multilayer adsorption on the monolayer surface also increases. This leads to a continuous increase in adsorbed layer thickness until the condensation pressure is reached. Type II isotherms are commonly observed in physisorption processes. A strong interaction between the adsorbate and adsorbent leads to a lower pressure requirement for monolayer formation (point B) and saturation [11]. Nitrogen adsorption on an iron catalyst at temperatures as low as −195 °C is an example of a type II isotherm [12]. Similarly, the adsorption of argon on sodalite nano zeolite at −196 °C also exhibits a Type II isotherm behavior [14], As illustrated in Figure 2.10b.

Type III isotherms are typically observed when a polar gas adsorbate adsorbs on a nonpolar surface with low adsorption capacity. At low partial pressures, the uptake is reduced due to repulsive interactions. However, when the partial pressure increases, the adsorption process becomes favorable for the polar gas adsorbate to adsorb on the nonpolar adsorbent surface [11]. Examples of Type III isotherms include the adsorption of bromine on silica gel at 79 °C [12], nitrogen adsorption on polyethylene [11], and nitrogen adsorption on 1,3,5-benzene dicarboxylate-based metal-organic framework at −196 °C [15] (see Figure 2.10c).

Type IV isotherms are associated with the adsorption of gas adsorbates on mesoporous materials. These isotherms are characterized by a hysteresis loop and a saturation plateau in the P/P_o range of 0.6–0.95. As the relative pressure increases, the hysteresis loop is caused by capillary condensation in mesopores. The isotherm shape resembles Type II at low relative pressure, forming a monolayer on a macroporous solid followed by multilayer adsorption at intermediate relative pressure. Capillary condensation occurs at high pressures, resulting in a steep increase in the adsorbed volume once all mesopores are filled. As the pressure continues to rise, adsorption progresses on the low external surfaces of the adsorbent [11]. Examples of Type IV isotherms include the adsorption of benzene on ferric oxide gel at 50 °C [12], nitrogen on $CdIn_2S_4$, nitrogen on TiO_2 [11], and nitrogen on silica gel at −196 °C [16] (see Figure 2.10d).

Type V isotherms represent the adsorption of an adsorbate in a multilayer form on mesoporous surfaces via physisorption. However, the process's adsorption heat is less than the adsorbate heat of liquefaction, indicating weak adsorbate-adsorbent interactions and low adsorption capacity. At low relative pressure, the adsorption process proceeds with a very slow uptake. As the relative pressure increases, monolayer completion occurs, followed by the formation of a multilayer, and capillary condensation begins. Once all of the mesoporous sites are filled with gas adsorbates, adsorption continues on the external surface until all available sites are occupied, and the isotherm becomes saturated [1]. Examples of Type V isotherms include water vapor adsorption on charcoal at 100 °C [12] and water adsorption on polycarbonate-derived AC at 599.85 °C [17] as illustrated in Figure 2.10e.

Type VI isotherms are associated with the adsorption of a gas adsorbate on the surface of an adsorbent with different types of adsorption sites, varying energies, and low porosity with a uniform surface. Gas adsorbates are built up in different partial pressures, step-wise, where each layer corresponds to a specific amount of gas adsorbed as a monolayer or successive layers, involving clusters with varying numbers of molecules. Methane adsorption on magnesium oxide, wherein each step represents the condensation of a single layer [1], and methane adsorption on magnesium oxide at −186.15 °C [18] (see Figure 2.10f).

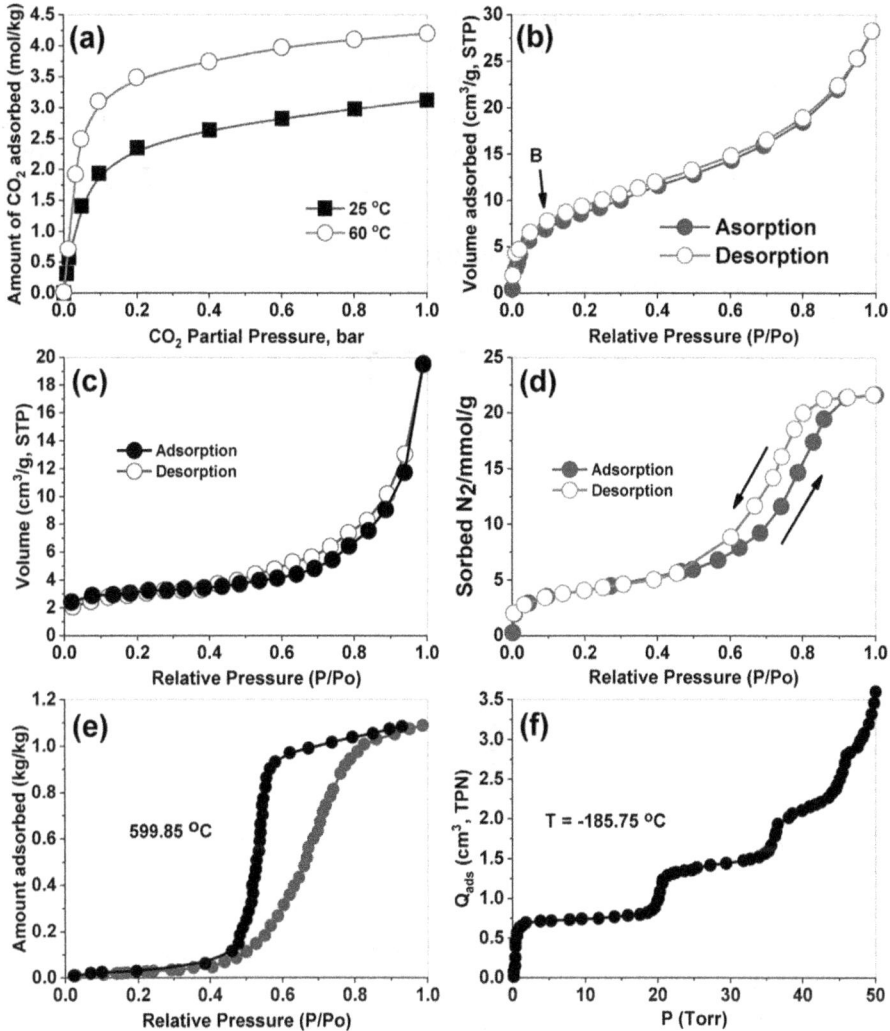

Figure 2.10: (a) Type I: CO_2 adsorption on Zeolite 13X at different temperatures (reproduced from ref. [13]). **(b) Type II:** Argon adsorption on sodalite nano-zeolite at −196.15 °C. Point B represents the completion of the first adsorbed monolayer and the start of multilayer adsorption (reproduced from ref. [14]). **(c) Type III:** N_2 adsorption on a 1,3,5-b3nzenetricarboxylate-based metal-organic framework at −196.15 °C (reproduced from ref. [15]). **(d) Type IV:** N_2 adsorption on silica gel at −196.15 °C. The upward and downward arrows represent the adsorption and desorption paths (reproduced from ref. [16]). **(e) Type V:** H_2O adsorption on polycarbonate-derived activated carbon at 599.85 °C represents the pyrolysis temperature used in synthesizing the activated carbon. Close and open symbols represent the adsorption and desorption paths (reproduced from ref. [17]). **(f) Type VI:** CH_4 adsorption on magnesium oxide at −186.15 °C. Close and open symbols represent the adsorption and desorption paths (reproduced from ref. [18]).

Chemisorption often entails a rapidly increasing curve that subsequently flattens. The initial rise signifies the strong tendency of the surface to bind gas molecules, while the leveling off may be attributed to the saturation of these forces through one or more of the three mechanisms. In contrast, physical adsorption is typically associated with an adsorption isotherm that features an increasingly positive slope as gas pressure escalates. Each incremental increase in gas pressure results in a more substantial increase in the amount of gas adsorbed, reaching its limit when the pressure equates to the vapor pressure of the adsorbed material. At this point, the adsorption isotherm ascends vertically as condensation occurs.

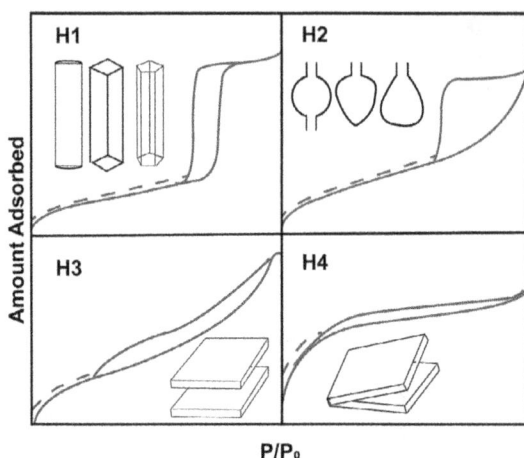

Figure 2.11: Representation of several types of hysteresis loops within the classification system provided by IUPAC (reproduced from ref. [20]).

The IUPAC acknowledges four types of hysteresis limits (Figure 2.11). **H1** hysteresis is characterized by parallel adsorption and desorption branches, indicative of relatively small pore sizes in non-connected mesopores, often observed in MCM-41 and SBA-15 materials. **H2** hysteresis typically appears as a triangular shape, generally caused by interconnected pores smaller than the body's entrance, referred to as "ink bottle"-like pores. **H3** hysteresis and **H4** hysteresis are attributed to the non-rigid pore structure between particle grains (**H3**) and flat plates (**H4**). The distinguishing feature of these two types of hysteresis is the absence of a defined flat plateau near $P/P_o = 1$, which typically results in the adsorption and desorption branches remaining separate without any significant overlap [19].

2.4.3.4 The Henry's adsorption isotherm

Henry's adsorption isotherm is effectively Henry's law applied to a two-dimensional solution instead of a bulk solution. It emerges as a special case (the low-pressure limit) of Langmuir's isotherm, as depicted in eq. (2.18) with $bP < 1$, where b is the amount of the surface adsorbate, P is the partial pressure of the adsorptive gas, V is surface coverage and k' is Henry's adsorption constant.

$$V = k'P \tag{2.18}$$

2.4.3.5 Freundlich isotherm

A Freundlich equation represents the Langmuir equation at intermediate values of θ in a convenient form. Zeldowitch, in his derivation of an adsorption isotherm for a heterogenous surface, showed that the Freundlich equation emerges from the conceptual framework that adsorption sites are distributed exponentially in relation to the heat of adsorption. Zeldowitch's derivation divides surface sites into several types, each possessing a unique, constant heat of adsorption. This means that the Langmuir model holds true for the sites within the same type, with no repulsion or any other mutual interaction affecting the adlayer. The fundamental equations of the form:

$$\frac{\theta_i}{1 - \theta_i} = b_i p \quad \text{and} \quad \theta = \frac{\sum_i \theta_i N_i}{\sum_i N_i} \tag{2.19}$$

where there are N_i sites of the ith kind, b is the amount of the surface adsorbate, P is the partial pressure of the adsorptive gas, θ is covered surface area and natural extensions of these equations, one arrives at

$$\ln \theta = \frac{RT}{A} \ln p + B \tag{2.20}$$

where A and B are constants, which can be rearranged to the form of the Freundlich equation, $\theta = k p^{1/n}$. Consequently, eq. (2.19) is likely to hold only for low coverages. The argument that the Freundlich equation predicts progressive coverage increases with rising pressure is erroneous, contrary to popular belief.

2.4.3.6 Temkin isotherm

The Temkin isotherm, also referred to as the Slygin-Frumkin isotherm, establishes a correlation between the covered surface area, θ, and the logarithm of the pressure, p:

$$\theta = A \ln Bp \tag{2.21}$$

The constant A is influenced by temperature, while constant B is linked to the heat of adsorption. The Temkin isotherm is derived from the assumption that the heat of adsorption decreases linearly with an increase in coverage.

2.4.3.7 The BET isotherm

The Langmuir isotherm expression is based on surface coverage with a single adsorbent layer. The isotherm curves for adsorptions that continue to increase, as in Figure 2.12, rather than flatten out after the initial adsorption stage, suggest a secondary adsorption stage. S. Brunauer, P.H. Emmett, and E. Teller described this process as an expression for the corresponding adsorption isotherm, known as the BET isotherm.

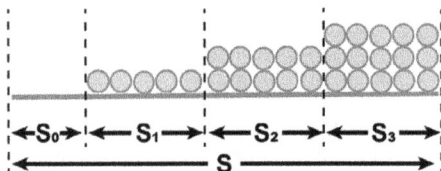

Figure 2.12: Schematic representation of an adsorbed film on an adsorbent surface.

To understand the BET isotherm expression, Figure 2.12 visualize the adsorption process. The initial layer of adsorbed molecules can adhere to the surface, as described by the Langmuir theory. In addition, adsorption can happen on top of the already adsorbed material. The surface can be represented as the sum of surfaces, S_0, S_1, S_2, S_3, etc., corresponding to the number of layers of molecules on the surface, with 0, 1, 2, 3, etc. layers, respectively. The complete adsorption process can be modeled using

– Adsorption onto solid surfaces.
– Adsorption onto surface sites already occupied by adsorbed molecules.

This approach incorporates an empirical constant, c, which represents the ratio of the strength of binding to a solid surface and that to adsorbed molecules. For adsorptions involving chemisorption, c has a value greater than 1, and in many cases, a value ranging from 100 to 1,000 is observed.

The theory yields an expression for the ratio of the gas volume that is adsorbed, V, to the volume that would produce a monomolecular layer, V_m. This ratio, V/V_m, is fractional and does not assume fractional values as it would in the adsorptions described by the Langmuir isotherm. The buildup of layers can continue until the adsorption reaches a stage where the surface can be considered "wet," when the pressure approaches the vapor pressure of the gas. To account for this limitation, it is convenient to work with the ratio of the actual pressure to the vapor pressure of the liquid adsorbate, represented by p/p^*. The mathematical expression for the BET isotherm is

$$\frac{V}{V_m} = \frac{cx}{(1-x)(1-x+cx)} \tag{2.22}$$

This expression can be rearranged to allow the experimental values of V versus P or V versus $x = P/P^*$ to be used to obtain values for V_m and c. We take the reciprocal of each side of the equation and then multiply both sides by V_m and by $x/(1 - x)$ to obtain

$$\frac{x}{1-x}\frac{1}{V} = \frac{1}{cV_m} + \frac{(c-1)x}{cV_m} \tag{2.23}$$

By plotting the left side of this equation against x, a straight line should be obtained, enabling us to determine the intercept on the $x = 0$ axis and the slope, which can be used to deduce the values of V_m and c. Estimated surface areas from such adsorption studies can be generally reliable but occasionally approximate. Table 2.2 illustrates the variation in the estimated area obtained from different gases, and Table 2.3 provides some typical surface areas for various adsorbents.

Table 2.2: Estimate the surface area of clean nickel films from the physical adsorption of different gases [3].

Gas	Area of adsorbed molecule (m²)	The amount adsorbed to give monolayer (cc)	Surface area of 1 g nickel film (m²)
Kr	14.6×10^{-20}	6.15	9.0
Kr*	16.6×10^{-20}	5.85	8.6
CH$_4$	15.7×10^{-20}	5.40	8.5
n-C$_4$H$_{10}$	24.5×10^{-20}	3.48	8.5

*On a different film.

Several nuances associated with the surface can lead to seemingly perplexing area values. For instance, fine pores or capillaries may allow one gas to permeate, while another gas with larger molecules finds them inaccessible. In this context, materials called molecular sieves, which consist of dehydrated zeolites, exhibit uniform-sized pores that serve as adsorbents. They have the ability to accept straight-chain hydrocarbon molecules while rejecting branched-chain molecules altogether.

Table 2.3: The volume of adsorbed nitrogen to form a monolayer and the surface areas of several catalysts (area covered by an adsorbed nitrogen molecule taken as 1.62×10^5 pm²) [3].

Material	Volume for monolayer, mL/g at 25 °C and 1 bar*	Surface area (m²/g)
Fused Cu catalyst	0.10	0.39
Fe, K$_2$O catalyst 930	0.16	0.61
Fe, Al$_2$O$_3$, K$_2$O catalyst 931	0.90	12.6
Fe, Al$_2$O$_3$ catalyst 954	3.2	12.6
Cr$_2$O$_3$ gel	60	230
Silica gel	130	510

*For this data, 1 bar = 1 atm.

2.4.3.8 The developments from Polanyi's adsorption theory

Working with microporous catalysts like zeolites, high-area carbons, and pillared clays sometimes requires reverting to one of the earlier adsorption theories proposed by Polanyi in 1914. According to Polanyi's treatment, the adsorption space close to a solid surface is characterized by a series of equipotential surfaces:

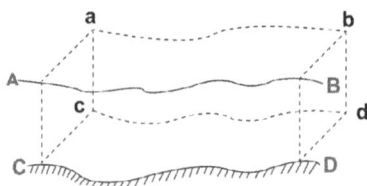

Figure 2.13: Diagrammatic representation of the adsorption space near an adsorbent. ABba is an equipotential surface that mirrors the topography of the adsorbent surface, CDdc (reproduced from ref. [21]).

In Figure 2.13, ABDC symbolizes a section of the adsorption space associated with the unit mass of a solid, while ABba depicts an equipotential surface. When the space (of volume W) between CDdc and ABba becomes filled with adsorbate, the equilibrium pressure being p, the adsorption potential at the surface ABba, symbolized by $\mathcal{E}$, is defined by

$$\mathcal{E} = RT \ln P_0/p \tag{2.24}$$

Polanyi visualized the adsorbate as being in intimate contact with the solid in a liquid state, where W is given by x/p, with x representing the adsorbed mass and p symbolizing the density of the liquid. Building upon this concept, Russian scientists Dubini and Redushkevich derived a new adsorption isotherm. The adsorption potential, resulting from dispersion and polar forces between the solid and adsorbed molecules, is independent on temperature but varies depending on the nature of the adsorbate and the adsorbent. Each of these forces is a function of the molecule's polarizability (α). Consequently, the adsorption potential ($\in$) of two distinct vapors at a given value of W on a specific solid will exhibit a constant ratio with respect to each other:

$$\frac{\in_1}{\in_2} + \frac{\alpha_1}{\alpha_2} = \beta \tag{2.25}$$

The constant β is called an affinity coefficient. If another adsorbate is chosen as the standard, eq. (2.25) transforms as follows:

$$\frac{\in}{\in_0} + \frac{\alpha}{\alpha_0} = \beta \tag{2.26}$$

The symbols with the suffix "0" represent the standard vapor, while those without this suffix relate to the other vapor. Dubinin and Radushkevich posited that the volume of the adsorption space can be expressed as a Gaussian function of the corresponding adsorption potential. For the standard vapor, this can be represented as

$$W = W_0 \exp\left(-A \in_0^2\right) \tag{2.27}$$

W_0 represents the total volume of all micropores, while A represents a constant that reflects the pore size distribution. Combining eqs. (2.26) and (2.27) leads to

$$W = W_0 \exp\left[-A\left(\frac{\in}{\beta}\right)^2\right] \qquad (2.28)$$

which, substituting from eq. (2.24), yields

$$\frac{x}{p} = W_0 \exp\left\{-\frac{A}{B^2}\left[\left(RT \ln\frac{p_0}{p}\right)\right]^2\right\} \qquad (2.29)$$

Which rearranges to

$$\ln x = \ln(W_0 p) - D \left[\ln\left(\frac{p_0}{p}\right)\right]^2 \qquad (2.30)$$

With D being $A(RT/\beta)$ [2], if we plot $\ln x$ against $(\ln(p_0/p))$ [2], we expect to obtain a straight line with a slope D and an intercept of $\ln(W_0\ p)$. Dubinin and others have shown that eq. (2.30) provides a handy method of assessing the micropore volume, W_0, as it produces a linear plot for a particular range of relative pressures, such as from $1 \times 10^{-5} \leq p/p0 \leq 0.2$ for benzene and other hydrocarbons on activated carbon (AC) samples.

2.4.3.9 The BJH method

A widely utilized method for determining the mesopore and macroporous size distribution of adsorbents is the Barrett, Joyner, and Halenda (BJH) analysis of the N_2 adsorption-desorption isotherm at 77 K. The BJH approach is utilized to derive distributions of pore volume and surface area from a comprehensive desorption or adsorption isotherm. The isotherm indicates the quantity of gas adsorbed onto a material at different relative pressures. Subsequently, the resulting pore size distribution showcases the amount of available pore volume and surface area within a particular range of radii, calculated based on the pressure ranges observed in the isotherm.

The initial BJH method was designed to determine pore size distributions using analysis of a desorption isotherm. This approach divides the desorbing gas into two groups: vaporizing cores (or inner capillaries) representing a significant portion of the total pore volume as a cylindrical tube concentric with the pore, followed by the gradual thinning of multi-molecular films exposed in pores whose cores have been emptied during previous desorption intervals. This forms the foundation for the BJH methods implemented by MicroActive, which include the Faas Correction, the Standard Correction, and the Kruk-Jaroniec-Sayari correction. This latter correction incorporates a minor modification to the Kelvin equation. Over the past two decades, the availability of well-defined model adsorbents in the mesopore range, combined with other forms of isotherm analysis aided by the density functional theory (DFT) has revealed that the BJH method tends to underestimate the pore size when the diameter falls below 10 nm.

2.4.3.10 The NLDFT method

A method for estimating micropore size was introduced by Horwath and Kawazoe and initially employed for predicting the pore sizes of microporous carbon. This approach has since been extended for application to other microporous materials. However, it is not suitable for in-ordered materials like zeolites and metal-organic frameworks (MOFs). Nonetheless, this method could be of interest when monitoring alterations in pore size due to post-synthesis techniques such as dealumination. Currently, the density functional theory (DFT) is the most commonly used mathematical modeling approach for constructing reference isotherms. Then, the advances in the understanding of DFT led to the recognition of Non-Local Density Functional Theory (NLDFT) approximations as superior for gas adsorption analysis, with the application of NLD. More advanced simulation methods, such as grand canonical Monte Carlo (GCMC), have also been employed.

The NLDFT method uses classical fluid density functional theory to construct the adsorption isotherms in ideal pore geometries, such as N_2 adsorption in slit-pores at 77 K. The pore size distribution (PSD) result is obtained by solving an adsorption integral equation using regularization techniques such as discrete Tikhonov regularization with non-negative least squares or a B-spline numerical technique. The two key components required for a meaningful evaluation of the pore size distribution are:

(a) A theoretical model (kernel) is employed to describe the relationship between adsorption isotherms and pore size as a function of pressure. This software incorporates models generated using 1D or 2D non-local density functional theory (NLDFT) for various types of materials.

(b) The pore size distribution is determined by a program that solves the adsorption integral equation (AIE), which can be achieved by fitting a model to experimental data. In SAIEUS, meaningful and stable results are obtained through a regularization technique that employs non-negativity constraints and optimizes the second derivative ("roughness") of the calculated distribution using the L-curve method.

The NLDFT method for pore size distribution determination can produce different results depending on the chosen gas and assumed pore geometry. However, recent research still commonly uses the simple slit-pore model for PSD determination, which is unlikely to be suitable for amorphous microporous materials with highly heterogeneous surfaces [19].

2.5 Chemical adsorption

2.5.1 Driving force to form a bond on an adsorbent surface

Heterogeneous catalytic processes rely on adsorption processes. The broken and formed bonds during adsorption reactions on a catalyst's surface will be discussed using basic molecular orbital theory and quantum mechanics. For instance, consider two H atoms, each having a positive nucleus and a negatively charged s electron. As the H atoms approach each other, their 2s orbitals will overlap, leading to the formation of two orbitals: a bonding orbital of low energy and an antibonding orbital of high energy, as depicted in Figure 2.14a. When the two H atoms overlap, two orbitals are created. The bonding orbital has a lower energy, while the antibonding orbital possesses a higher energy. The bonding orbital contains two electrons, making it stronger due to the presence of two nuclei. Additionally, an antibonding orbital is formed, which contains an empty electron.

Furthermore, interactions between two atoms can be classified into two categories: (i) major overlap: this type of overlap will stabilize the molecule and lead to an increased splitting gap between the bonding and antibonding orbital energies. (ii) Minor overlap: the splitting between bonding and antibonding orbital energies will be reduced, as depicted in Figure 2.14b. The strength of the adsorbate bond depends on the electronic filling of the bonding and antibonding orbitals.

When the electronic orbits of a bonding orbital are completely filled and no electrons are present in the antibonding orbital, a strong adsorbate bond forms on the surface. However, this is not ideal for catalysis, as breaking such a strong bond can be difficult. A more favorable situation occurs when the bonding orbital is filled with electrons, while some electrons reside or half-filled in the antibonding orbital, creating an intermediate bond. This scenario is more favorable for catalytic processes. If

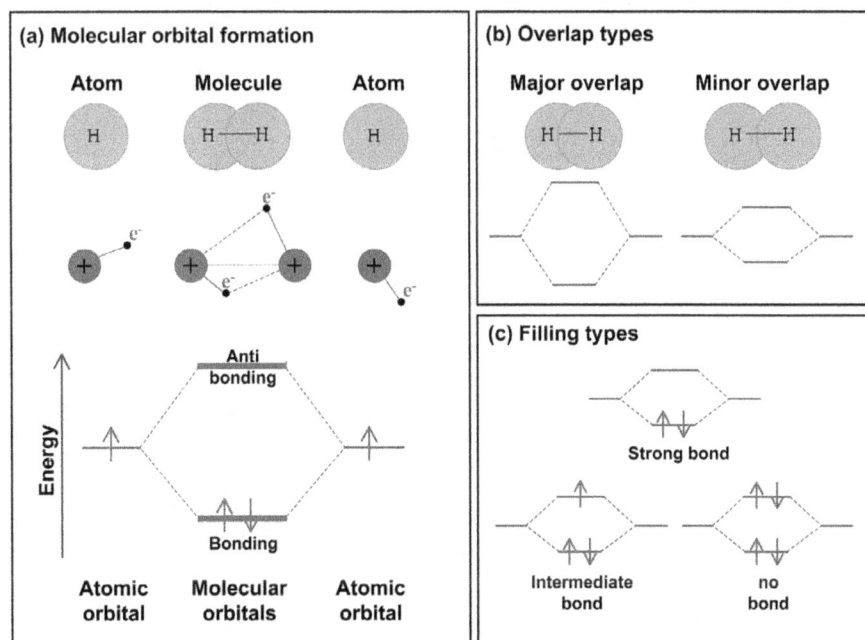

Figure 2.14: The basic principles of molecular orbitals: **(a)** The formation of bonding and antibonding orbitals, **(b)** overlap types of atoms, and **(c)** filling types (reproduced from ref. [22]).

all available orbits, including both bonding and antibonding orbits, are filled with electrons, no bonds will form, as seen in Figure 2.14c.

In the case of adsorbent surfaces made of metal (or d-metals) (Figure 2.15a), the overlap of atomic orbitals produces bands. The highest occupied orbit is known as the Fermi level, denoted as E_F. Electrons in the Fermi level of a metal can be promoted to a vacuum level, E_{vac}, located outside the metal's surface. The minimum energy necessary to remove an electron from a metal to the vacuum level is called the work function.

Consider a free atom being adsorbed onto the surface of a transition metal (or d-metal). A bonding orbital and an antibonding orbital will be formed. In the case of a reactive metal adsorbent, such as chromium ($3d^5$), strong chemisorption occurs when the bonding orbital is filled with electrons. The antibonding orbital remains empty, as shown in Figure 2.15b. Another example is when the d-metal has less reactivity, such as rhodium ($4d^8$). In this case, an intermediate chemisorption bond strength forms with moderate strength, wherein the electrons fill the bonding orbital and partially occupy the antibonding orbital, as illustrated in Figure 2.15c.

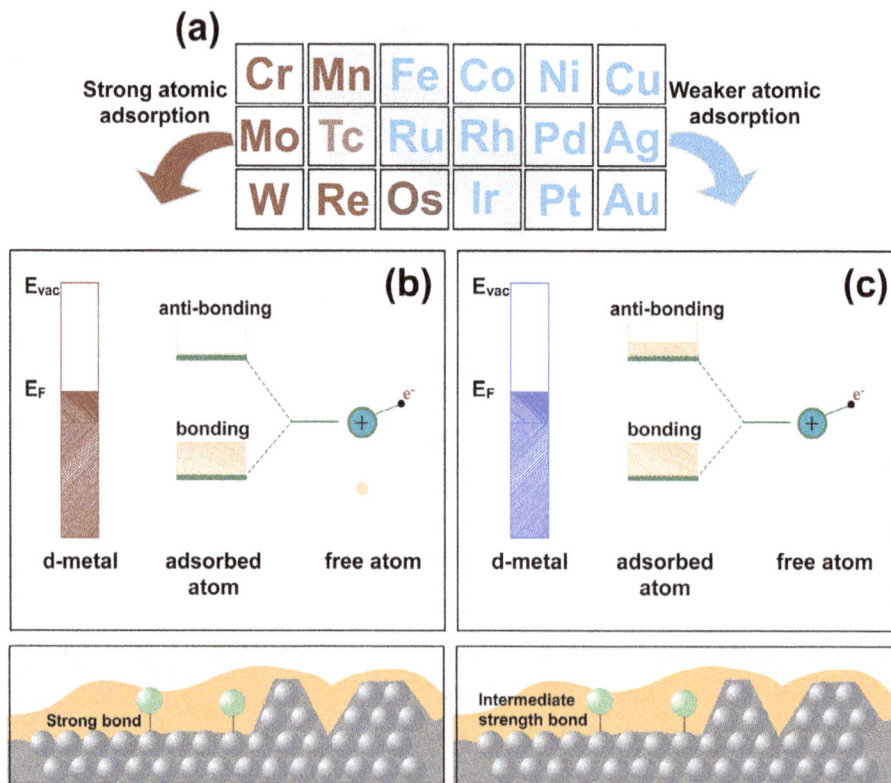

Figure 2.15: **(a)** Elements of d-metal in the periodic table for the weak or strong adsorption, **(b)** a strong chemisorption bond, and **(c)** a weak chemisorption bond (E_{vac} represents the energy of the vacuum level and EF represents the energy of the Fermi level; reproduced from ref. [22]).

On the other hand, when a molecule is adsorbed onto a metal surface (d-metal), the bonding and antibonding orbitals of the free molecule will interact with the metal. The bonding orbital of the free molecule (σ) will be filled by electrons to create a new orbital called the adsorption orbital with the metal. The resulting bonding adsorption orbital will be fully occupied with electrons, while the antibonding adsorption orbital, being only partially filled, will lead to a moderate bond strength. Similarly, the antibonding orbital of the free molecule (σ^*) will also create an adsorption orbital with the surface metal. When the antibonding orbital of the free molecule is involved in adsorption, this weakens the internuclear bonds of the free molecule. In cases of dissociative adsorption, the bonding orbital derived from the antibonding orbital of the free molecule (σ^*) should be completely filled, leading to the breakdown of the free molecule and its complete dissociation.

2.6 Different synthesis strategies

Heterogeneous catalysis plays a crucial role in modern society. Catalyst reactions occurring on the surface are heavily influenced by the atoms within the catalyst layer, making it crucial for specific catalyst surfaces to interact with reactants efficiently for optimal efficiency. This emphasis on catalyst surfaces has led to an increased interest in nanoscale catalysts, known as nanoparticles, due to their high surface-area-to-volume ratios, which contribute to enhanced reactivity and optical absorptivity. The active catalytic sites are primarily located at low-coordinated atoms including steps, edges, and corners. Thus, if nanomaterial (NMs) surfaces are structured in a manner that exposes a high density of low-coordinated atoms such as steps, edges, and corners, they can exhibit exceptional catalytic activity. However, the crystallization thermodynamics of these high-energy NM surfaces tends to decrease rapidly under normal growth thermodynamic conditions, leading to their disappearance. Consequently, synthesizing NMs with high-energy surfaces presents a significant challenge in catalytic and controlled shape synthesis. Additionally, high-energy surface nanocrystals offer a promising basis for understanding fundamental principles of surface science and heterogeneous catalysis, such as the relationship between nanostructures and properties, which holds the key to the design of high-activity and practical catalysts for energy conversion and storage [23].

Heterogeneous catalysts materials containing diverse types including metallic, non-metallics and metal oxides. Typical industrial heterogeneous catalysts often consist of inorganic materials like metals and metal oxides, which provide high thermal stability, essential for many industrial applications, as shown in Table 2.4.

Table 2.4: Some examples of large-scale industrial processes [24].

Catalyst	Reaction
Ni, Pd, Pt (transition metals as powders or on support)	C = C bond hydrogenation: e.g., olefin + H_2 → paraffin
Cr_2O_3 (metal oxides)	
Cu, Ni, Pt (transition metals)	C = O bond hydrogenation: e.g., acetone + H_2 → isopropanol
Pd, Pt (noble metals)	Complete oxidation of hydrocarbons: oxidation of CO
Fe (metal supported and promoted with alkali metals)	Haber process: $3H_2 + N_2$ → $2NH_3$
Ni (transition metal)	Methanation: $CO + 3H_2$ → $CH_4 + H_2O$ Steam reforming: $CH_4 + H_2O$ → $3H_2 + CO$
Fe or Co supported and promoted with alkali metals	Fischer-Tropsch reaction: $CO + H_2$ → paraffins + olefins + H_2O + CO_2 + oxygen – containing organic compounds
Cu/ZnO-Al_2O_3	Synthesis of methanol: $CO + 2H_2$ → CH_3OH
Re + Pt/μ-Al_2O_3 Re + Pt/Y-Al_2O_3 and promoted with chloride	Paraffin dehydrogenation, isomerization, and dehydrocyclization. Naphtha reforming: n – heptane → toluene + $4H_2$
SiO_2-Al_2O_2, zeolites (solid acids)	Paraffin cracking and isomerization
Y-Al_2O_3	Alcohol → olefin + H_2O
Pd supported on acidic zeolite	Paraffin hydrocracking
Metal oxide-supported complexes of Cr, Ti, or Zr	Olefin polymerization: ethylene → polyethylene
Metal oxide supported oxides of W or Re	Olefin metathesis: 2 propylene → ethylene + butene
V_2O_5 or Pt	$2SO_2 + O_2$ → $2SO_3$
Ag on inert support promoted by alkali metals	Ethylene + $\frac{1}{2}O_2$ → ethylene oxide(with $CO_2 + H_2O$)
V_2O_5 on metal oxide support	Naphthalene + $\frac{9}{2}O_2$ → phthalic anhydride + $2CO_2 + 2H_2$
	o – Xylene + $3O_2$ → phthalic anhydride
Bismuth molybdate, uranium antimonate	Propylene + $\frac{1}{2}O_2$ → pacrolein
	Propylene + $\frac{3}{2}O_2 + NH_3$ → acrylonitrile + $3H_2O$
Mixed oxides of Fe and Mo	$CH_3OH + O_2$ → formaldehyde(with $CO_2 + H_2O$)

Table 2.4 (continued)

Catalyst	Reaction
Fe_2O_4 or metal sulfides	Water gas shift reaction: $H_2O + CO \rightarrow H_2 + CO_2$
Ni (metal)	Hydrogenation of vegetable oil
Cu (metal)	Oxidation of alcohols
$\left\{ \begin{array}{l} Co - Mo/\ \Upsilon - Al_2O_2\ (\text{sulfide}) \\ Ni - Mo/\Upsilon - Al_2O_3(\text{sulfide}) \\ Ni - W/\Upsilon - Al_2O_3\ (\text{sulfide}) \end{array} \right\}$	Olefin hydrogenation, aromatic hydrogenation, hydrodesulfurization, and hydrodenitrogenation

2.7 Methodologies for nanomaterials (NMs) synthesis

Various synthesis and processing techniques have been developed to produce nanostructured materials. There are two primary methods in NM synthesis: **(1)** the top-down approach and **(2)** the bottom-up approach. The top-down approach involves starting with larger bulk building units that are then reduced in size systematically, bit-by-bit, leading to finer nanostructures. Conversely, the bottom-up approach involves assembling or coalescing smaller building blocks, such as molecules or nanoclusters, to create suitable units with required properties, as depicted in Figure 2.16 [25, 26]. Furthermore, nanostructured materials can be synthesized through one or a combination of physical, chemical, biological, or more methods. Notably, the preparation method can significantly impact surface properties, influencing the material's performance and application potential.

2.8 Top-down approach

Large materials are typically deconstructed through physical and chemical processes in the top-down approach to nanostructured materials synthesis. The top-down approach using physical methods relies on photons, electrons, and ions. Conversely, the chemical top-down strategy involves chemical reactions triggered by chemical etching agents and the application of heat [27]. The top-down approach is challenging and expensive for large-area assembly, it has the advantage of avoiding the use of harmful chemicals or biological species.

2.8.1 Pulsed laser ablation in liquid (PLAL) method

The laser ablation method is a photo-thermal process that involves irradiating a bulk material target with a laser, resulting in the absorption of laser energy. For nanosecond (ns) lasers, this leads to either the heating (melt flow) or vaporization of the target. In

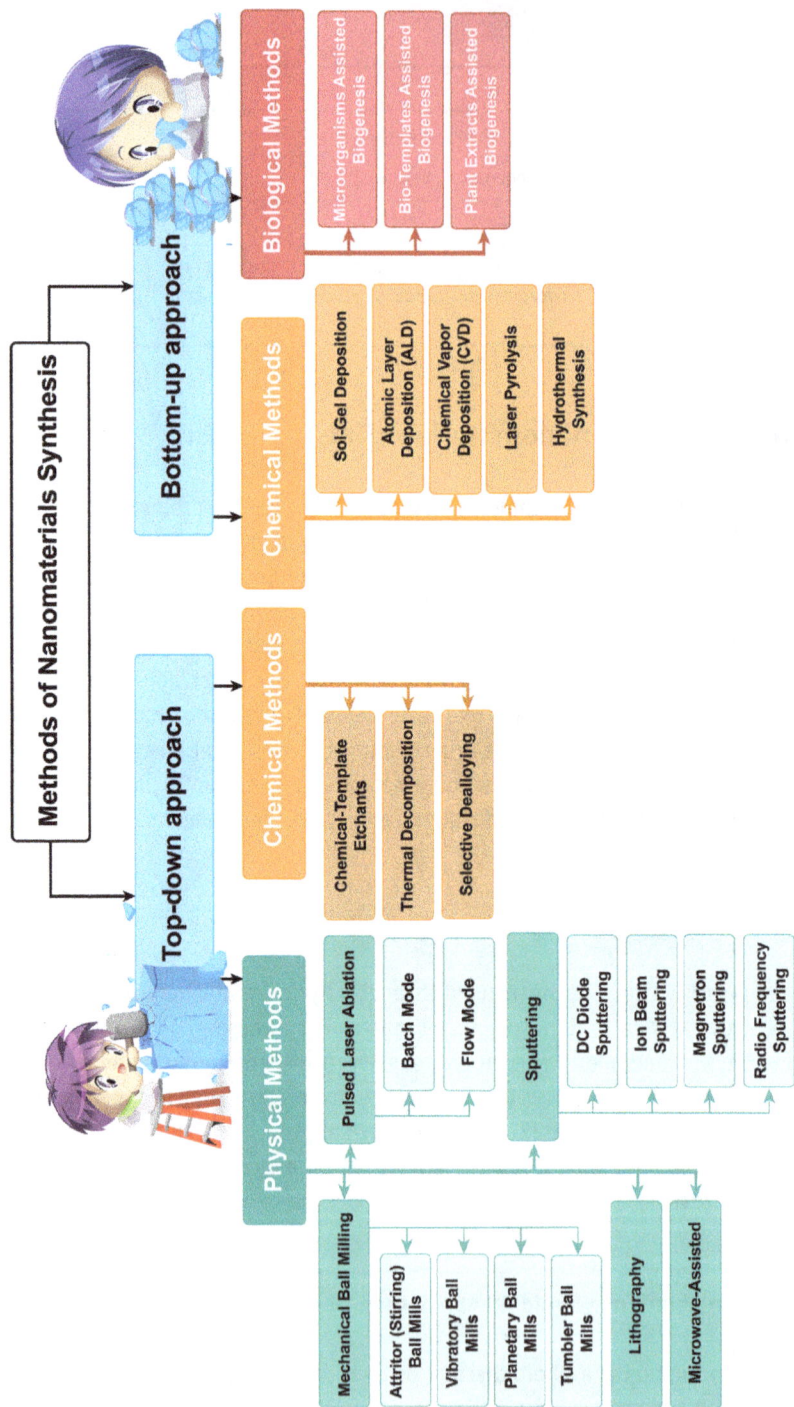

Figure 2.16: Comprehensive summary of different types of synthesis methods to produce a variety of nanoparticles.

contrast, pico- (ps) and femtosecond (fs) lasers can cause Columbic phase explosions of the target [28–31]. There are two modes of PLAL: (a) batch mode (Figure 2.17a) and (b) flow mode (Figure 2.17b). The primary differences between these modes are the motion-flow or lack-flow of a liquid environment thereof during ablation [32]. The standard setup for PLAL includes a pulsed laser, beam delivery optics, and a container that holds both the target material and liquid [31]. Both modes of PLAL have their advantages. The batch mode is primarily utilized for research purposes, while the flow mode has recently gained popularity in industry due to its higher NP yield and colloidal density [33, 34]. PLAL presents several benefits, including lower equipment cost, ease of the method, no requirement for specific conditions like a vacuum, and the ability to control material properties through laser parameter adjustments and liquid media selection. However, the plasma shielding effect can hinder the interaction between incident laser photons and the target, resulting in a decrease in ablation efficiency if the plasma density surpasses the critical density of the incident lasers photons. To reduce the shielding effect, researchers have employed two strategies: (a) rotating either the target or the ablation container [32, 35] or (b) laser scanning [36–38]. The PLAL process can be divided into four distinct stages: (i) generating a shock wave on the target, producing plasma plumes and elevating the temperature of the target to its liquid or vapor phase; (ii) formation of cavitation bubbles within the liquid, encapsulating small particles, ions, and atoms, along with their associated interactions; (iii) releasing the ablated material from the cavitation bubbles into the liquid medium, generating a second shock wave. subsequently, (iv) the aging and growth of nanostructures occur, as depicted in Figure 2.17c. The PLAL method stands as one of the most reliable and customizable ways to fabricate nanostructures due to its capability to fuse various materials using laser energy. The key challenge lies in selecting the optimal laser processing parameters and the liquid medium employed water, acetone, ethanol, isopropyl alcohol, etc. to achieve required outcomes.

The PLAL method can synthesize various materials including nanoparticles, metal oxides, metals, carbon nanotubes (CNTs), nanowires, zeolites, composites, semiconductor quantum dots, and core-shell nanostructures [31] (see Table 2.5). For instance, Ferman et al. [39] used the PLAL method to synthesize zinc oxide (ZnO) in double-distilled water (DDW), utilizing a zinc plate immersed in a glass container along with 1 mL of deionized (DI) water. The Nd: YAG (1,064 nm) laser was employed, delivering 500 and 600 mJ per 100 pulses to obtain two ZnO NMs types. Meanwhile, Hesabizadeh et al. [40] synthesized spherical copper oxide nanostructures (CuO/Cu_2O) by PLAL. A copper target consisting of 15 copper beads was submerged in 25 mL of DI water. The experiment was conducted with an Nd: YAG laser operating at 1,064 nm with a repetition rate of 5.1 kHz, a pulse energy of 2 mJ/pulse, and a 30-min irradiation period. Griaznova et al. [41] prepared Fe-Au core-satellite nanoparticles using the PLAL method. The synthesis process involved two steps. First, a 50 mL solution of 1 mM NaCl was prepared and then the Au target was added. The solution was exposed for 40 min to a femtosecond pulsed laser radiate from Yb: KGW (1,030 nm wavelength, 100 kHz repetition, and 250 fs pulse duration 3×10^{-5} mJ pulse energy), with 3 mm beam diameter and focused by a 100 mm F-theta

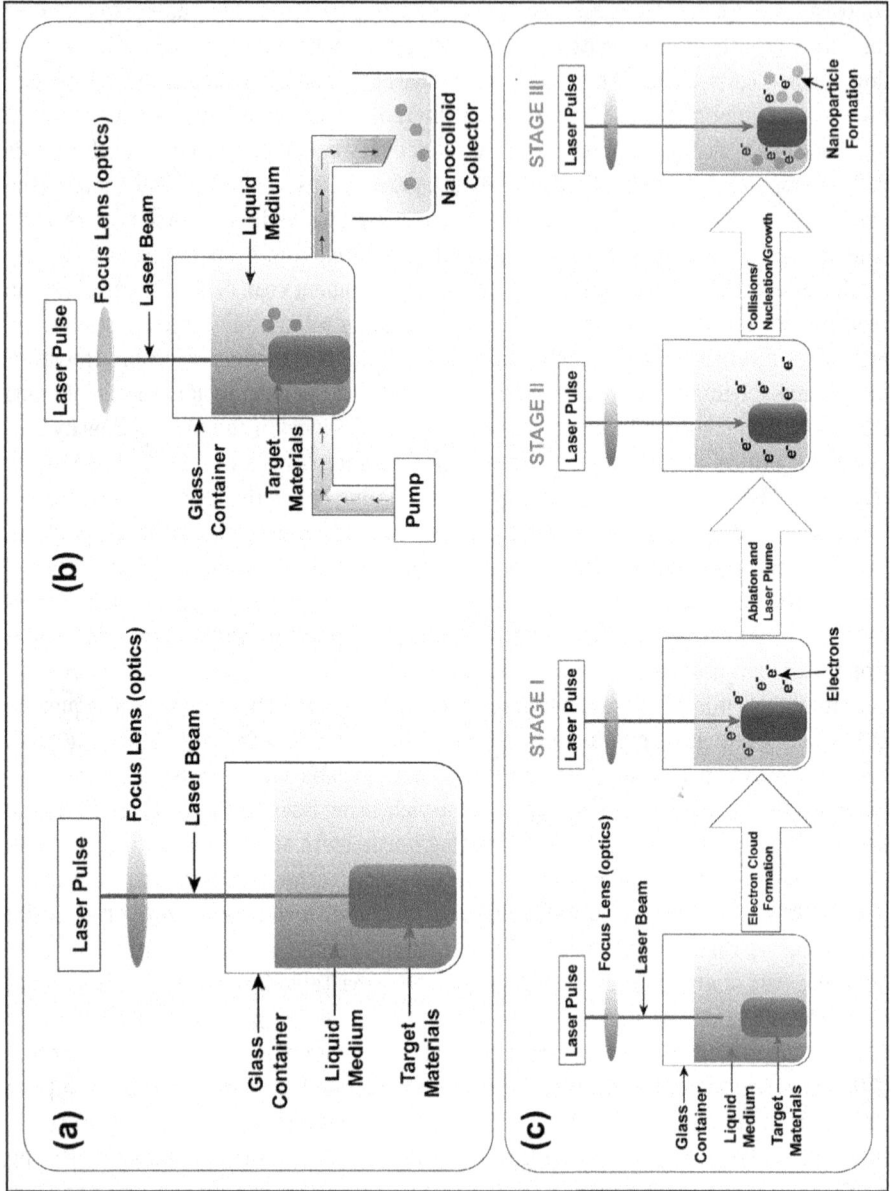

Figure 2.17: Schematic representation of pulsed laser ablation in liquid (PLAL): (a) batch mode PLAL, (b) flow mode PLAL, and (c) illustration of the PLAL process in batch mode showing the three main stages of PLAL.

lens on the surface of a target. Then, the colloidal solution was centrifuged for 15 min. The solution was subjected to radiation from a femtosecond for 40 min pulsed Yb:KGW laser with a wavelength of 1,030 nm, a repetition rate of 100 kHz, a pulse duration of 250 fs, and a pulse energy of 3×10^{-5} mJ. The laser beam had a diameter of 3 mm and was focused by a 100 mm F-theta lens onto the target's surface. Then the colloidal solution was centrifuged for 15 min. Subsequently, Fe metal in the form of NPs was mixed with the Au target in solution and underwent a similar process as described in the initial step for 15 min. The non-reacted Au was removed using the magnetic field provided by an NdFeB cylindric magnet for 5 min, attracting only the magnetic fraction of NPs to the bottom of the vessel while non-magnetic NPs remained suspended in the solution. To stabilize the obtained Fe-Au NPs, 1 mL of 50 mM polyacrylic acid was added, and the solution was heated to 80 °C. Finally, the NPs were incubated for 15 min at 80 °C and washed three times with water to remove any unbound polymer via centrifugation for 10 min. In addition, Almessiere et al. [42] attempted the synthesis of a nanocomposite using the PLAL method. They synthesized nanocomposites of $SrGd_{0.03}Fe_{11.97}O_{19}$ (SGFO-hard): $CoTm_{0.01}Fe_{1.98}O_4$ (CTFO-soft) with varying ratios (1:1, 1:1.5, 1:2, and 1:2.5) through the PLAL method. Both SGFO and CTFO were first synthesized via the sol-gel auto-combustion method. Subsequently, the exact amounts of SGFO and CTFO were transferred to glass beakers and mixed with 20 mL of DI water (with a pH of ~7). The two suspensions were separately subjected to 20 min of ultrasonication and then combined under 1 h of ultrasonic homogenization. Next, the mixed solution was subjected to PLAL while stirring for 40 min. The irradiation source consisted of a Nd:YAG nanosecond laser with a pulse duration of 9 ns, a repetition rate of 10 Hz, a wavelength of 532 nm, and a pulse energy of 3,000 mJ. Finally, the SGFO: CTFO hard-soft ferrite nanocomposite was dried for 2 h at 120 °C. Further effort by Deng et al. [43] synthesized thin films of silicate-1 (SIl-1), mordenite (MOR), and UTD-18 using the PLAL method on stainless steel foil. They prepared all silicate target materials step-by-step. To create the targets, they utilized a gel composition consisting of $1SiO_2$:$165H_2O$:$0.32TPAOH$ (tetra propylammonium hydroxide). Fumed silica served as the silica source, while TPAOH acted as the template. The solution of these two components with DI water under stirring to form a gel. Following the formation of the gel, it was aged at room temperature for 60 min while being stirred. Subsequently, the gel was heated to 170 °C and maintained under static conditions for 24 h to obtain silicalite-1. Following this, silicate-1 and MOR (containing ferrocene) were loaded into a 23 mL Teflon-lined Parr reactor and heated to approximately 130 °C, above the melting point of ferrocene, for 24 h. Once cooled to room temperature, the resulting solid, which was light brown in color, was washed with dichloromethane multiple times until the filtrate became colorless. Afterward, the material was dried at room temperature for 24 h. Additionally, UTD-18 was prepared by utilizing a mixture composition of 0.13 [$(MeCp)_2Co$]OH or [$(EtCp)_2Co$]OH: $0.1NaOH$: $1SiO_2$: $60H_2O$ under hydrothermal conditions at 175 °C for 6 days in Teflon-lined stainless-steel autoclave. For the PLAL method, the three silicate materials – silicalite-1, MOR, and UTD-18 – were pressed, forming a target pallet approximately 2.5 cm in diameter. Prior to placing the target in the chamber, the substrate – a stainless-steel foil with a thickness of 50 μm – was heated to 250 °C with a quartz lamp, and the background O_2 pressure was maintained at 250 mTorr. Then, the targets were placed in a controlled atmosphere chamber, positioned above the stainless-steel foil substrate. A 14-nanosecond pulse laser was employed, delivering 50–100 mJ of energy per pulse at a repetition rate of 10 Hz for a total duration of 30 min.

Table 2.5: Different types of nanomaterial synthesized using the PLAL method.

NMs	Precursor (target)	Liquid media	Laser type	Wave. (nm)[a]	P. Ene. (mJ)[b]	Irr.Time (min)[c]
ZnO	Zinc plate	Double-distilled water (DDW)	Nd: YAG (ns)[d]	1,064	500 and 600	– [39]
CuO/Cu_2O	Copper beads	Deionized (DI) water	Nd: YAG (ns)[d]	1,064	2	30 [40]
Fe-Au core-satellite	Au metal + Fe metal	NaCl solution	Yb: KGW 250 (fs)[e]	1,030	3×10^{-5}	15 [41]
SGFO: CTFO	Powder from sol-gel	Deionized (DI) water	Nd: YAG 9 (ns)[d]	532	3,000	40 [42]
SIL-1:MOR: UTD-18/Aelfa	Silicates-1 + Mordenite + UTD-18	–	Nd: YAG 14 (ns)[d]		50–100	30 [43]

[a]Wavelength (nm).
[b]Pulse energy (mJ).
[c]Irradiation time (min).
[d]Nanosecond lasers (ns).
[e]Femtosecond lasers (fs).

2.8.2 Mechanical milling method

Mechanical milling is a non-equilibrium physical manipulation method that involves solid-solid interactions between different elemental powders, milled together in an inert atmosphere with high energy to form homogenous mixed powders with the same composition as the constituents. This process is repeated until the desired particle sizes and dimensional properties are achieved. The primary objectives of mechanical milling are twofold: (i) to reduce the size of particles and (ii) to transform the particle phase into a new one, facilitating the mixing and synthesis of NPs [31]. The primary advantage of mechanical milling is its simplicity, making it a cost-effective choice suited for large-scale productions. However, this method has limitations in producing high-quality end products, and increased ball milling temperature can lead to a temporary halt in the process to cool down the temperature [44]. The method involves interactions between balls and powder within the milling apparatus. Various powder characteristics, such as composition, phase equilibria, and stress levels during the milling process and classification (e.g., ductile-ductile, ductile-brittle, and brittle-brittle systems) [45], affect the subsequent mixing, pressing between balls, deformation, and breakdown of particles, ultimately influencing the final properties of the milled powder products. Two possible scenarios occur during ball milling: (i) the balls may move layer-by-layer within the milling chamber, using the chamber's surface for support or (ii) the balls can fall to the bottom of the chamber without using the surface and potentially influence the powder already present below the balls. The ball milling process involves four stages: (i) the welding stage, where irregularly shaped powder particles weld to each other upon impact under the influence of high forces from the balls; (ii) the squeezing stage, where the morphology of welded particles transforms into a sheet that becomes increasingly thinner with continued impacts; (iii) fracturing stage, in which further impacts lead to fragmentation into the required size range suitable for metal and in the final stage (iv) a dynamic stable stage, further fragmentation and intermixing occur to create a stable size range [46, 47]. Several critical parameters control the properties of powder produced during mechanical milling including (i) the raw materials used, (ii) the type of milling equipment, and (iii) the reaction conditions [48]. A number of parameters influence the energy imparted by the balls in mechanical milling: (i) milling type, (ii) milling speed, (iii) ball size, (iv) milling conditions (wet or dry), (v) temperature, (vi) material density, and (vii) processing time. The majority of milling balls are constructed from hard materials such as stainless steel or ceramics [46]. Utilizing ceramic balls is often preferred as the kinetic energy increases proportionally with their mass. If a denser ball is used, the movement of the balls can be minimized, reducing the potential for collisions. The duration of the process is closely monitored to ensure that the final product maintains a consistent chemical composition ratio compared to the initial stock powder [31]. Researchers categorize mechanical mills based on several criteria such as rotation speed [44], activity [46], and mechanism [31]. When considering rotation speed, ball mills can be divided into two main groups: (i) high-energy ball mills and (ii) low-energy ball mills [44]. High-energy ball mills encompass various types such as attritor ball mills, vibratory (shaker) ball mills, and planetary ball mills. Low-energy ball mills, on the other hand, include tumbler ball mills [31, 44]. Based on their activity, ball milling can be categorized into two primary groups: (i) direct and (ii) indirect. The direct process involves the physical contact of rollers or mechanical shafts with the particles and the direct transfer of kinetic energy (such as in attritor ball mills). In the indirect process, kinetic energy is initially transferred to the body of the apparatus and then redistributed to the grinding media. Devices that fall under this category include (i) Attritor mills, (ii) vibratory ball mills, (iii) planetary ball mills, and (iv) tumbler ball mills.

(i) **Attritor (or stirring) ball mills** are efficient and straightforward. They grind the starting materials using free-moving beads set in motion by a stirring mechanism. Milling occurs due to the agitation of an agitator, which consists of a vertical rotating central shaft with horizontal arms known as impellers. The rotating speed of the agitator falls within a range of 75–500 revolutions per minute [44] (see Figure 2.18a).

(ii) **Vibratory ball mills:** This powerful technique is used to produce high-quality fine disperse powders. The operating principle of the vibratory mill is based on the intense vibratory motion and motivation of the grinding bodies due to the use of inertia and centrifugal forces rather than gravitational forces. In this process, the rotation of the vibrator shaft and the movement of the mill body cause the grinding balls to move under the eccentricity or radius of the carrier. The energy transfer from the grinding charge is carried out through the milling apparatus. The influence of inertia, centrifugal forces, and cyclic loads causes the balls within the body to follow a complex trajectory, compressing against the walls of the drum and colliding with one another and the particles of the crushed material. In this process, the balls break, crush, and grind the particles [49], as illustrated in Figure 2.18b.

(iii) **Planetary ball mills:** The method depends on the rotational speeds of the grinding jars, and the support disc can be independently adjusted. By altering the gear ratio, the movement and trajectory of the grinding balls can be controlled, enabling them to strike horizontally against the inner wall of the grinding jar, approach tangentially, or roll over its inner wall. An array of rotational speeds and combinations of pressure, friction, and impact can be readily achieved [49], as depicted in Figure 2.18c.

(iv) **Tumbler (or horizontal) ball mills:** The method encompasses a revolving circle partially packed with steel balls around its longitudinal axis. The device's diameter primarily influences the effectiveness of this technique, as larger diameters result in greater fall height and, thus, higher energy transfer to the balls [46] (see Figure 2.18d).

Mechanical milling has been employed to manufacture amorphous, nanocrystalline, metal-organic frameworks, alloys, and metal/nanometal composite materials. For instance, Hong et al. [50] prepared aluminum flake powders (Al-FP) by tumbler ball milling from aluminum foil scraps. The aluminum foil scraps, characterized by a thickness of 6.5 mm, a width of 6 mm, lengths between 50 and 250 mm, and a purity of 99.4 wt.%, were first cut into chips with a size below 5 mm using rotary blades. As previously pointed out, 126 g of aluminum foil scrap, 3 wt.% oleic acid as a milling agent, and 126 g of mineral spirits were loaded into the jar. Millings were performed at varying times, ranging up to 22 h, at different rotational speeds corresponding to cascade and cataracting modes. In a similar manner, as outlined by Li et al. [51], amorphous cobalt vandates oxide (CVD) was synthesized through planetary ball milling. CoO/MnO and V_2O_5 were mixed together and ball-milled for 10 h. Subsequently, 10 wt.% of commercial graphite was added to the mixture and milled for an additional 5 h to enhance electronic conductivity and promote dispersion. A weight ratio of 28:1 was maintained between the balls,

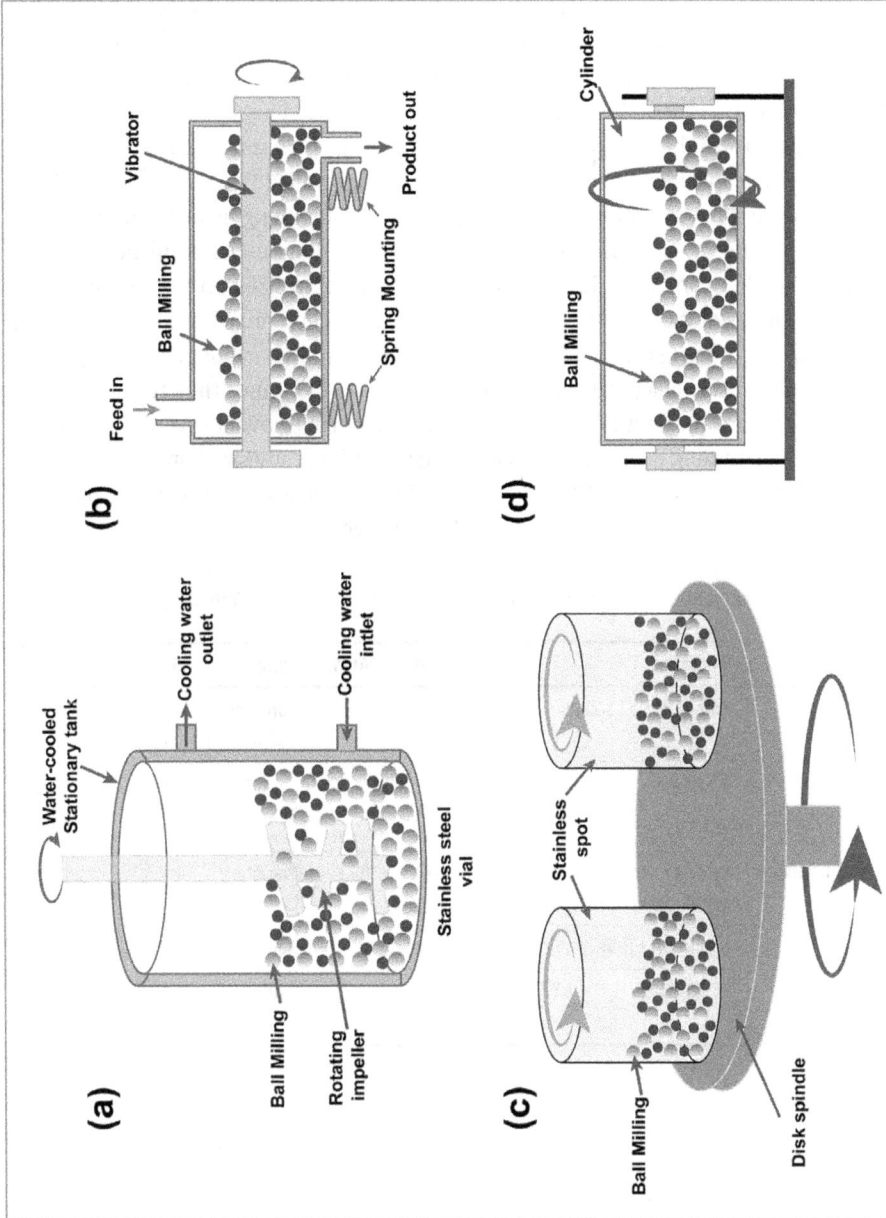

Figure 2.18: Schematic of ball milling processes: (a) attritor (stirring), (b) vibratory ball mixer, (c) planetary ball mixer, and (d) tumbler (horizontal) ball mixer.

and the powder used in the ball-milling process for 7-mm diameter balls was utilized. A 10 min pause was taken after each operation hour during the ball-milling procedure. To achieve high crystallinity of CVO and MVO, the CVO and MVO materials were treated at 400 °C for 2 h at a rate of 2 °C/min in an Ar environment. Liu et al. [52] synthesized sulfur-CNT composites through planetary wet ball milling. They mixed a precursor sulfur/CNTs composite, including 2 g of sulfur and 2 g of multi-walled CNT, which was added to a 250 mL agate tank and thoroughly mixed through ball-milling in 10 mL of ethanol or 10 mL of chloroform. The ball-milling process was carried out under ambient conditions, at a speed of 300 rpm, for 3 h. Almotairy et al. [47] synthesized an ultra-strong Al/SiC composite material using various ball milling speeds. They mixed aluminum powder, which was 98% pure and had an average size of 30 μm, with nanosized β-SiC (β-silicon carbide), which was 95% pure and had an average size of 60 nm. They further incorporated citric acid into the mixture as a processing control to avoid severe cold welding during the ball-milling process. Qu et al. [53] successfully synthesized an amorphous Zn-Al-layered double hydroxide. This was achieved by subjecting the mixture to planetary ball-milling for 2 h at a rotational speed of 600 rpm (see Table 2.6).

Table 2.6: Different types of nanomaterial synthesized using the ball milling method.

NMs	Precursor	Dry/wet	Atm.[a]	Type	B:P ratio[b]	Time (h)
Al-FP	Aluminum foil scraps	Oleic acid	–	Tumbler ball milling	–	22 [50]
MVD	$MnO + V_2O_5$	–	Ar gas	Planetary ball milling	28:1	10 [51]
CVO	$CoO + V_2O_5$	–				
S/CNT composite	Sulfur + multi-walled carbon nanotubes	Ethanol/ chloroform	Air	Planetary ball milling	–	3 [52]
Al/SiC	Aluminum + β-SiC	Stearic acid	–	Ball milling	15:1	2 [47]
Zn-Al LDH	$Zn_4CO_3(OH)_6.H_2O + Al(OH)_3$	–	–	Planetary ball milling	–	2 [53]

[a]An atmosphere.
[b]Ball: powder ratio.

2.8.3 Sputtering fabrication method

Sputtering is a non-thermal evaporation process where a target's surface is bombarded by energized ions of a plasma (a partially ionized gas). This interaction leads to the physical ejection of atoms from the target's surface, which then deposit onto a substrate. The efficiency of this process relies on the energy of the incident ions. To generate a plasma, relatively high pressures (within the range of 10^{-1}–10^{-3} mbar) are necessary. Initiating at lower pressure before introducing Argon is crucial to prevent contamination due to residual gases [31]. Figure 2.19 illustrates the basic schematic of sputter deposition. The target, made up of source material, is positioned opposite the substrate. The target (cathode) is typically subjected to a negative voltage in the 0.2–5 kilovolts (kV) range. The substrate can be secured and transformed into an anode. In this scenario, an inert working gas (most commonly Argon) is introduced into the chamber. A continuous pump regulates the pressure during the inflow (usually within a range of 0.1–10 Pa). Positively charged ions pick up momentum and fly toward the target, sputtering target atoms and then traveling to the vapor phase to be deposited onto the substrate. The sputtering process also produces free electrons propelled away from the cathode by collisions within the vapor phase, generating more ions and electrons and perpetuating the sputtering process. This chain reaction creates a neutral mix of energized electrons and ions between the target and substrate, known as plasma discharge or glow discharge since excited ions often emit visible photons as they return to their ground state. Consequently, the scientific principles behind ion-surface interactions consist of (i) the impact of an ion on a solid surface; in this instance, the solid surface is referred to as the target and (ii) the reflection and subsequent neutralization of the ion upon impact. This scattering phenomenon forms the foundation for an analytical technique called ion scattering spectroscopy, thereby enabling (iii) the ion to penetrate the target. The term for this process is ion implantation, which is extensively used in integrated circuits for selective doping of silicon wafers. It facilitates precise control over the depth and quantity of impurity atoms and (iv) may induce structural alterations in the target material. These modifications can arise from fundamental vacancies, industrial defects, and significant lattice imperfections (defects), capable of instigating changes in target composition, and (v) ultimately, the ion collision can trigger a succession of collisions among target atoms, culminating in the ejection of one of these atoms. This ejection process is referred to as sputtering [54]. This sputtering phenomenon boasts numerous advantages, with the concentrated deposited film equating to that of the original material.

Sputtering came to light during numerous attempts to deposit a thin film onto a liquid substrate. At that time, the researchers faced constraints in recording nanoparticle deposition due to limited technological advancements. Nonetheless, for the first time, they observed instances of magnetic colloidal solution sputtering onto a liquid surface and subsequently confirmed metallic colloidal sputtering onto silicon oil. However, it was established that silicone oil was inadequate in stabilizing colloids. Consequently, it became necessary to opt for another liquid medium. Various techniques of sputtering exist (i) DC diode sputtering, (ii) ion beam sputtering, (iii) magnetron sputtering (MS), and (iv) radio frequency (RF-diode) sputtering.

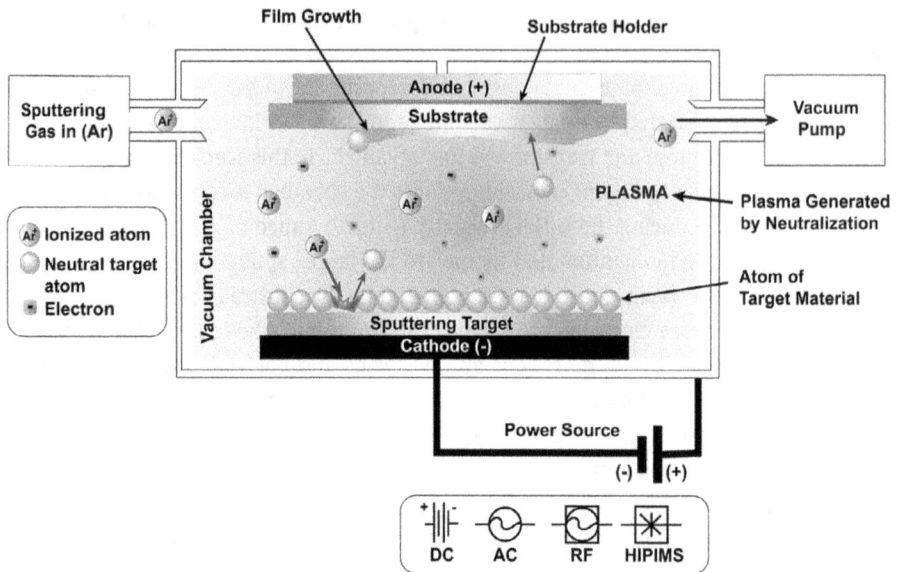

Figure 2.19: Schematic representation of the setup of the sputtering process.

(i) **DC diode sputtering**: This system consists of a tandem of planar electrodes. One electrode function as the target (the cathode), while the other, serving as the anode, is grounded, as depicted in Figure 2.19. The exterior face of the cathode is cooled with water. The sputtering chamber is maintained at a pre-determined pressure (P) and a few millivolts (direct current, within the range of 500–1,000 V) are applied between the target and the substrate, thereby instigating a glow discharge. Positive Argon ions within the plasma accelerate toward the cathode, causing the target atoms to sputter and deposit thin films onto the substrate. The cathode (target) in a DC diode system must be conductive since insulating surfaces can generate a charged surface that hinders the impact of ions on the target's surface and interrupts the sputtering process. The primary drawback of this type of sputtering is the relatively slow sputtering rate achieved, as a limited number of Argon ions are generated, and it only applies to electrically conductive materials.

(ii) **Ion beam sputtering:** This mechanism involves the generation of a relatively concentrated ion beam in a separate chamber, achieved through the use of Kaufman-type ion guns. The ions are then extracted and directed into the sputtering chamber, wherein they strike the target under controlled vacuum conditions, thereby enabling a high bombardment rate. This methodology permits a reduced pressure within the deposition chamber, ensuring that the mean free path of the film-forming entities greatly exceeds the distance between the target and the substrate. Although seldom employed for deposition purposes, it is commonly used to achieve depth-profile measurements through photoelectron or Auger electron spectroscopy and to prepare transmission electron microscopy samples through ion milling techniques [54].

(iii) **Magnetron sputtering (MS)**: This system incorporates magnets placed behind the target to create curved field lines parallel to the target's surface. The magnetic field can be achieved using permanent magnets, electromagnets, or a combination of both. Free electrons are accelerated by the target's negative potential and exhibit a spiral motion along the magnetic field lines as a result of the Lorentz force. Employing a magnetic field close to the target enhances the ionization rate, thereby reducing pressure and plasma density and ultimately increasing deposition rates.

(iv) **Radiofrequency (RF) sputtering**: This technique utilizes a radiofrequency (RF) electric field oscillating at approximately 50 kHz, which prompts the electrons to oscillate and gain sufficient energy for ionization. This results in the formation of a plasma that does not depend on electrons sourced from the target. The electrons in this RF plasma possess much higher mobility than the ions, leading to a greater influx of electrons than ions onto the target surface. Using an insulating target, however, necessitates equalized currents of impinging electrons and ions, resulting in a negative self-bias on the target. This DC self-bias accelerates ions toward the target, facilitating the sputtering process. It is mainly employed in the deposition of insulating (e.g., aluminum oxide or boron nitride) and semiconductor materials, resulting in reduced substrate heating and higher sputtering rates compared to DC sputtering. Nevertheless, this method does have certain drawbacks, such as lower coating rates, expensive RF generators, and potential unevenness in the plasma density, especially for large rectangular cathodes ("1 m), inconsistent plasma density leading to uneven layer thickness distribution in deposited films, and DC triode sputtering.

In recent times, many researchers have utilized sputtering technology to form a thin film coating over a desired material substrate, surface modification, and the synthesis of electrocatalysts. For instance, Poelman et al. [55] developed a new type of vanadia/titania catalyst coating on an amorphous SiO_2 support using DC MS for oxidative dehydrogenation of propane. Orozco-Montes et al. [56] deposited platinum (Pt) onto a Vulcan 72 powder to form a carbon-supported platinum nanoparticle catalyst using the MS method at various pressures (1.0, 4.0, and 9.0 Pa). Xu et al. [57] synthesized superhard AlCrTiVZr high-entropy alloy nitride films on a Si wafer employing high-power impulse MS (HiPIMS) at differing nitrogen flow rates (FN) ranging from 0 to 20 cm. Zhao et al. [58] coated the Zr-4 substrate with AlTiCrNiTa using the RF sputtering method to investigate its corrosion tendencies. Shen et al. [59] examined the oxidation resistance of an $(Al_{0.34}Cr_{0.22}Nb_{0.11}Si_{0.11}Ti_{0.22})_{50}N_{50}$ thin film coated on a Si wafer by the MS method. Li et al. [60] synthesized a FeAlCuCrCoMn coating on a quartz glass wafer to examine its mechanical behavior. Zhang et al. [61] fabricated an amorphous AlFeCoNiCuZrV coating using the DC MS process. The sputtering was conducted under argon and oxygen atmospheres to explore the influence of the oxygen flow ratio on high-entropy alloy thin films' mechanical properties and microstructure.

2.8.4 Lithography fabrication method

Lithographic technology constitutes a system used to generate thin films with functional structures and adjustable dimensions. It is a cost-effective method that is classified into two types: (i) conventional lithography and (ii) unconventional lithography approaches (see Figure 2.20). Conventional lithography entails transferring particular patterns to thin, responsive materials known as resistors that are deposited on a substrate through internal reactions between particles and the substrate's surface (photons and electrons). The pattern on the resist can be transferred by (a) direct or (b) indirect methods. The direct (parallel) writing method is a maskless technique, such as electron beam lithography (EBL) and laser writing lithography (LWL). Electron beam lithography (Figure 2.21a), transfers the required pattern pixel-by-pixel, which adds accuracy but is unfortunately slower. Laser writing lithography involves the direct dispensing of a precursor solution of nano-particles on a glass substrate, followed by laser-induced photoreduction and nanoparticle patterning. The dispensing and laser patterning processes are repeated until the final 3D structure is obtained. The excess solution of reduced nanoparticles is then flushed away with solvent (such as ethanol) [62] as illustrated in Figure 2.21b. In contrast to indirect (serial) writing, optical lithography operates by transferring the required pattern to the substrate through a mask containing the pattern in a single step, see Figure 2.21c. It typically includes a light source and a mask with the intended pattern. Masks can be positioned at varied distances from the light source, and different contact and non-contact modes are available. Light can pass through areas without metal or be reflected after striking a metallic region, with masks classified as binary based on this property. The pattern acts as an intermediary, with the resist covering the surface substrate. By arranging the interaction between the light and resist, specific molecule resistance configurations can be achieved.

Contrarily, the second option, the unconventional lithography method, is nanoimprint lithography (NIL), which can be divided into UV-based NIL, thermo NIL, laser-assisted NIL (LANI), and electrochemical NIL. Figure 2.21d represent the UV-based NIL operates at ambient temperature and low pressure, requiring surface substrates coated with UV-curable liquid resist. The resistance material is exposed to UV radiation and solidified under UV radiation. An optical transparent mold form is then pressed into the substrate to extract patterns. Transparent molds offer high-precision optical alignment. The second technique involves the thermo-imprint method as shown in Figure 2.21e, include a mold made from rigid or flexible polymer material. During the printing process, the required pattern is formed on the polymer using the mold. This pattern is then transferred to the substrate after separation from the mold

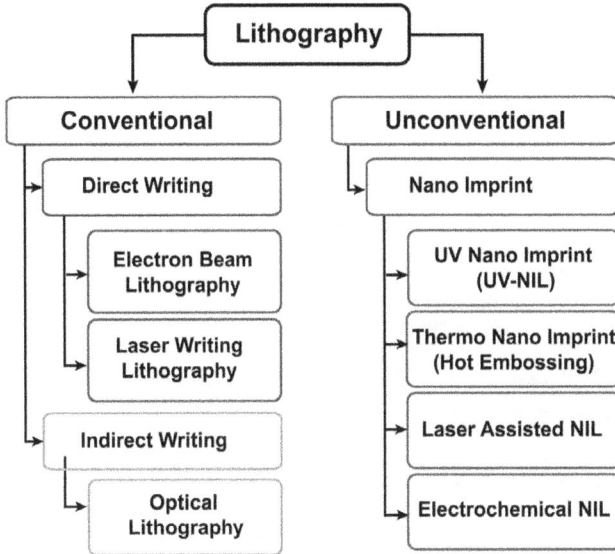

Figure 2.20: Schematic representation of different lithography techniques.

through the application of thermo-NIL and hot embossing [31]. Furthermore, laser-assisted nanoimprint (LA-NIL) is depicted in Figure 2.21f, is a resistless technology that avoids the need for etching. This technique involves exposing a single excimer laser pulse to a transparent quartz mold, thereby melting a thin layer of the silicon substrate surface. The liquid layer is subsequently engraved by the quartz mold. Upon cooling the substrate, the mold is then released. With LA-NIL, various nanostructures with a resolution of less than 10 nm can be imprinted onto silicon wafers in less than 250 ns. The last technique is electrochemical nanoimprinting (EC-NIL) is in Figure 2.21g, a resistless approach that utilizes an conductive mold. A voltage is applied between the mold and the target substrate after which, once the mold surface comes into contact with the substrate, current flows between them causing the anodic oxidation of the substrate surface, and subsequently, the substrate is etched to achieve the nanostructure [31, 63]. These techniques offer a broad scope for creating 3D structures at the microscale, facilitating the development of high architectural complexity, making them valuable in various fields such as tissue engineering, cell biology, drug delivery, photonics, plasmonics, metamaterials, and micro-electro-mechanical system. For instance, Ryu et al. [64] synthesized highly aligned single-walled carbon nanotubes on quartz, while Danilevicius et al. [65] developed a hybrid organic-inorganic sol-gel based on silicon-zirconium oxides using direct laser writing.

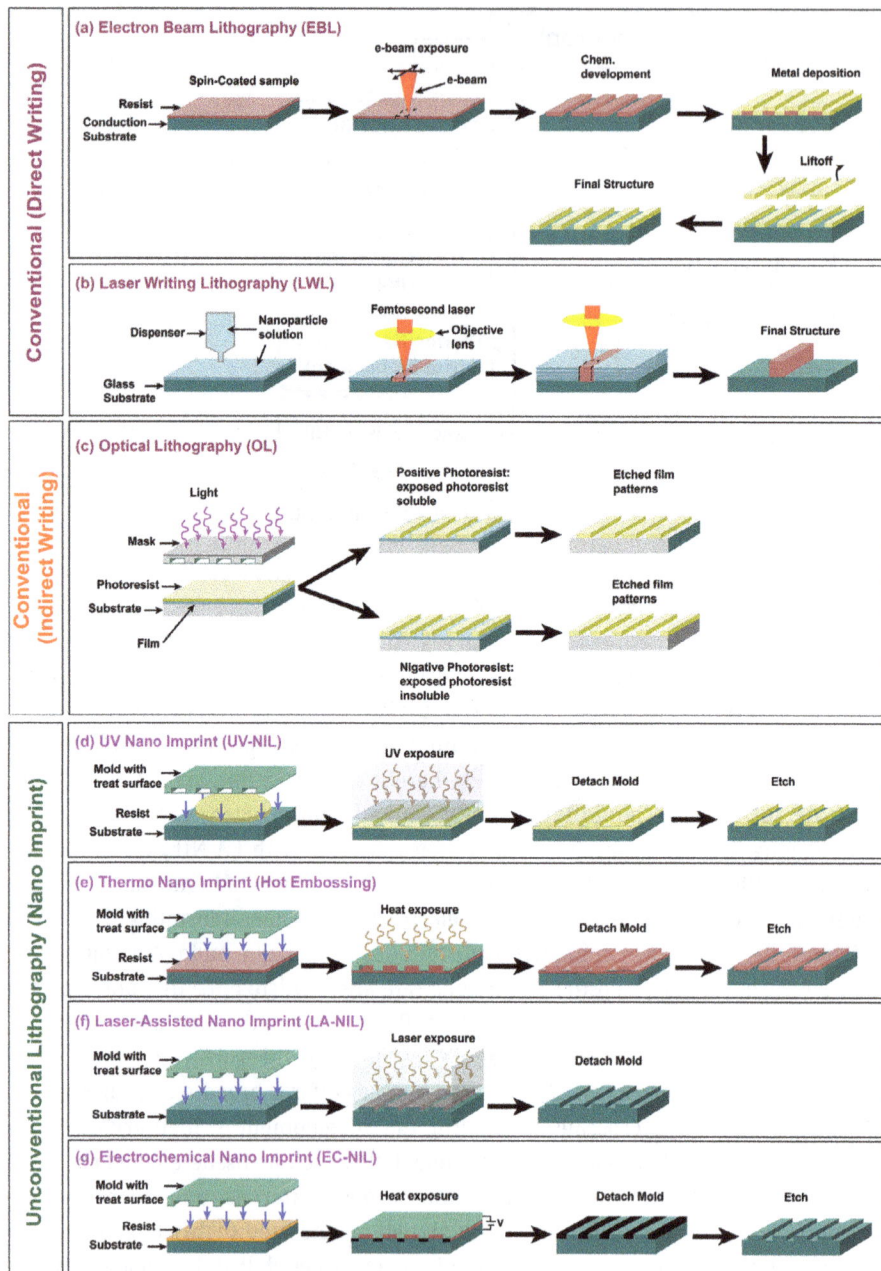

Figure 2.21: Schematic representation of different conventional and unconventional lithography techniques. The conventional includes (a) electron beam lithography (EBL), (b) laser writing lithography (LWL), and (c) optical lithography (OL). The unconventional includes (d) UV Nano imprint (UV-NIL), (e) thermos nanoimprint (hot embossing), (f) laser-assisted nanoimprint (LA-NIL), and (g) electrochemical nanoimprint (EC-NIL) (reproduced from ref. [63]).

2.8.5 Microwave-assisted fabrication method

Microwaves (MW), which are electromagnetic waves with wavelengths in the range of 0.3–300 mm, have a frequency corresponding to the rotational energy of molecules, making them an efficient and low-energy heating mechanism. Compared to infrared, visible, and ultraviolet sources, microwaves possess lower energy. Microwave irradiation has proven to be an effective method for synthesizing NMs as it rapidly and homogeneously heats precursor materials through the combined action of electric and magnetic fields. This results in molecule collisions and frictions, which facilitate rapid crystal growth and the formation of crystallites with a narrow size distribution [66]. Microwave heating is also a clean method as it produces less waste and is efficient in terms of production speed. It heats faster than conventional heating methods and transfers energy through radiative conductivity and convection, as shown in Table 2.7. Materials react to microwave heating in different ways: reflect, absorb, or transmit microwaves, depending on their chemical properties. Polar liquids absorb microwaves easily, converting the energy into heat due to the combination of dipolar interaction and ionic conductivity [67]. However, one of the drawbacks of microwave heating is that the material's core tends to be at higher temperatures than its outside surface, leading to the risk of explosion [68]. Common liquid media used for NP synthesis by MW method include (i) DI water, (ii) alcohol, (iii) DMF, and (iv) ethylene glycol, which are ideal due to their high dielectric losses. When exposed to microwaves, material ions and dipoles interact with electromagnetic forces, causing the material's ions to rotate in alignment with the magnetic field. The rotation is delayed due to the friction of surrounding ions and dipoles, resulting in a relaxation time when ions and dipoles lose energy and uniformly distribute heat throughout the material. Microwave heating depends on both (i) dielectric parameters and (ii) penetration depth. It is critical to consider dielectric losses during NP synthesis, as they determine how efficiently microwave energy is converted into heat, which is necessary for NP production.

This process is typically carried out in a specially designed oven that includes a flask, a solution bowl, and a magnetron amplifier. An optic fiber thermometer is connected to the chamber to measure temperature data which is then transmitted to a control system. Additionally, a magnetic stirrer is placed at the base of the oven to agitate the solution, which is often coated with Teflon. The flask is also connected to an electric supply of 300 to 1,100 W and a condenser that mixes the precursor with certain additives. A schematic of such a system is depicted in Figure 2.22. The apparatus can be modified to include a light system that simultaneously measures the size distribution and concentration of NPs during synthesis as a quality control measure. A dynamic light scattering system works by using a laser to release light that is scattered by the NPs, which is then captured by a detector and related to the NPs' size and motion. Moreover, a fluorescent temperature probe can be added to record the target material's temperature under microwave irradiation. Numerous physicochemical phenomena occur during microwave-assisted NP synthesis, including non-thermal

Table 2.7: Different between conventional heating and microwave-assisted heating [69].

Conventional heating	Microwave-assisted heating
– The heating or thermal reaction starts from the surface of the material only.	– The heating or thermal reaction starts uniformly and simultaneously from the surface to the bulk of the material.
– The heat transfer requires physical contact between the material surface and the vessel.	– Physical contact between the material surface and the vessel is not required.
– Heating takes place from an electric or thermal source.	– Heating takes place by microwaves.
– The heating of material involves a thermal conduction mechanism.	– Heating of material involves dielectric polarization of material.
– Uniform heat is applied to all the material systems.	– Heating of a specific system is possible.
– Low heating rate.	– Higher heating rate.

effects such as hotspots, hot surfaces, superheating, and the effect of MW irradiation on the apparatus and the solvent [31]. The microwave-assisted method can be employed for both extraction and synthesis using reactors. Microwave-assisted extraction (MAE) is utilized for extracting bioactive compounds from plants, and there are different types of MAE methods: (i) MAE under vacuum, where the extraction process is performed in a vacuum environment; (ii) solvent-free microwave extraction, which eliminates the need for solvents and is suitable for extracting both volatile and non-volatile compounds; and (iii) MAE combined with ultrasound to enhance the extraction process.

Conversely, the microwave-assisted synthesized reaction using a Teflon reactor (hydrothermal) was conducted by Onwudiwe et al. [70] who synthesized PbS nanostructures with the microwave irradiation of a single source. They dispersed a 0.5 mM Pb (II) Di thiocarbamate in 20 mL of ethylenediamine and sonicated for 20 min to obtain a homogeneous dispersion. The resulting mixture was then transferred to a Teflon-sealed and heated in a microwave oven (multiwave 3,000 microwave sample system, power = 800 W) for 20 min, with stirring after cooling the solution to room temperature. The dark brown products formed were isolated by centrifugation, washed several times with absolute ethanol and dried at 50 °C overnight. Mustafa et al. [71] synthesized the nickel cobalt phosphate ($NiCoPO_4$) composite powder using microwave-assisted techniques. They simultaneously investigated two factors, spin speed and spin time, to evaluate their impact on the specific capacities of $NiCoPO_4$ composites. The solution consisted of 100 mM $NiCl_2.6H_2O$, 100 mM $CoCl_2.6H_2O$, and 0.2 M Na_2HPO_4 was mixed and transferred into a Teflon container. This solution underwent a microwave-assisted hydrothermal reaction at a temperature = 124 °C, heating

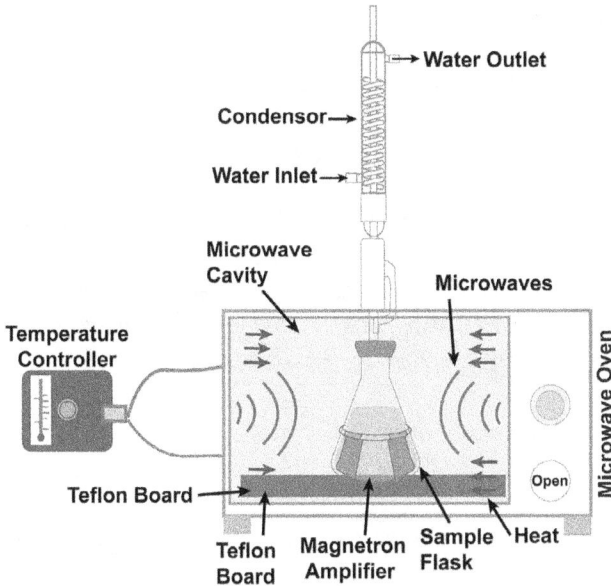

Figure 2.22: Schematic representation of the microwave-assisted method.

rate = 4 °C/min, holding time = 10.5 min and a holding time of 10.5 min. The ratio of nickel to cobalt was set at 1:1. The obtained $NiCoPO_4$ composite powder was dried in an oven (80 °C for 12 h) before being dispersed in ethanol (100 mg/mL) and spin-coated onto the pre-cleaned ITO substrate. Kubiak et al. [72] synthesized yttrium-doped TiO_2 through a comparative analysis of the hydrothermal method using both traditional and microwave techniques. They prepared a 1 wt.% titanium (IV) chloride solution in distilled water with an ice-water bath. Then, 100 cm^3 of the $TiCl_4$ solution was transferred to an IKA reactor, and 1 g of urea was added with continuous stirring for 15 min. The solution was then transferred to a hydrothermal or microwave reactor for heat treatment. In the hydrothermal approach, the heat treatment parameters were set to $T = 200$ °C and $t = 12$ h. On the other hand, for the microwave treatment, the settings were $T = 200$ °C, $t = 1$ min, and $P = 300$ W. After the process, the reactor was allowed to cool down to room temperature, and then the obtained substance was washed three times with DI water and dried at 60 °C for 6 h. Many scientific studies have explored the advantages of utilizing microwaves over traditional heating. However, the limited application of this technology in industries is attributed to the substantial capital costs and lower energy conversion efficiency compared to conventional heating methods.

2.8.6 Chemical-template etching method

Chemical templated etching is a method of engraving the substrate that uses a template with the required nanoscale pattern adhered to the substrate surface to direct the chemical etching. The template material is then removed, leaving behind a permanently etched pattern crafted on the substrate surface. The etching process can be performed either dry or wet (see Table 2.8). The wet chemical etching method has been significant in creating patterns; however, the isotropic nature of wet etching reduced the sizes and surface topographies, resulting in a discrepancy between the feature size defined by the mask and that replicated on the substrate (Figure 2.23). Consequently, the use of dry etching (anisotropic) or combining two methods may be necessary to overcome these challenges.

Table 2.8: The difference between wet etching and dry etching [73].

	Wet etching	Dry etching
Definition	Wet etching involves using liquid chemicals or etchants to remove material from a surface substrate.	Dry etching involves removing material using gases or plasmas in a vacuum chamber.
Materials used	Acid, bases, and other solvents. Common etchants include hydrofluoric acid, nitric acid, and potassium hydroxide.	Gases like sulfur hexafluoride, carbon tetrafluoride, and oxygen.
Anisotropy level	Generally isotropic, meaning it etches uniformly in all directions.	Highly anisotropic, allowing for precise control over the shape and profile of the etch.
Selectivity	Varies widely depending on the etchant and material: sometimes less selective.	High selectivity, allowing for precise removal of specific materials without affecting others.
Equipment	Requires less sophisticated equipment. It often consists of baths or tanks for immersion.	Requires more complex and expensive equipment like plasma etch or reactive ion etch (RIE) systems.
Uniformity	It may have less uniformity across a wafer due to fluid dynamics and agitation requirements.	Controlled gas flow and reaction conditions offer high uniformity across the wafer surface.
Surface damage	There is a lower risk of inducing surface damage due to the absence of high-energy ions.	A higher risk of surface damage is due to bombardment by high-energy ions and radicals.
Throughput	High throughput for batch processing of water.	Lower throughput: typically processes wafers one at a time.
Environmental impact	Generates liquid waste, which requires proper disposal and treatment.	Generates gaseous by-products: some may be toxic and require treatment before disposal.

Anisotropic etchings, also known as dry etchings, use specialized technology to remove the mask or template through physical means, like the impact of ions that cause the ejection of the template or mask from the substrate. Alternatively, chemical reactions can convert the substrate material into volatile reaction products that can be removed through pumping. Dry etch technology encompasses several commonly employed methods that can be categorized as follows: (a) chemical etching (including isotropic radial etching), (b) physical etching (reactive ion etching, sputter etching, ion milling, and ion beam-assisted etching), and (c) a combination of both methods (reactive ion beam etching). All dry etch techniques are carried out under vacuum conditions, with pressure influencing, to some degree, the specific phenomena occurring in each process.

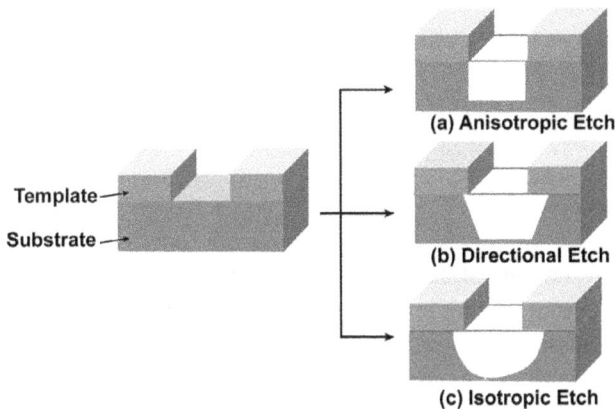

(a) Anisotropic Etch

(b) Directional Etch

(c) Isotropic Etch

Template

Substrate

Figure 2.23: Schematic representation of the character wet etches processes: (a) anisotropic etch, (b) directional etch, and (c) the isotropic etch.

For instance, in Figure 2.24, a silicon substrate was fabricated using a specific block-copolymer template, combining vinyl pyridine as the core and styrene-based polymers in a toluene-rich solution. After spin-coating, the amphiphilic polymer micelles self-assembled into a pseudohexagonal array on the substrate surface to form the template. Nanoscale models were then selectively engraved into the base substrate using an aqueous hydrofluoric acid (HF) solution. When silicon surfaces are etched using fluoride-based solutions, only the poly(4-vinyl pyridine) core is affected, as the protonation of the pyridine group by HF causes selective placement of fluoride ions within the micellar core. Upon removal of the polymer model through ultrasonic treatment, in toluene, a silicon etch pit array is formed, which can then be functionalized with additional materials, like gold nanoparticles, to create intricate architectures [74].

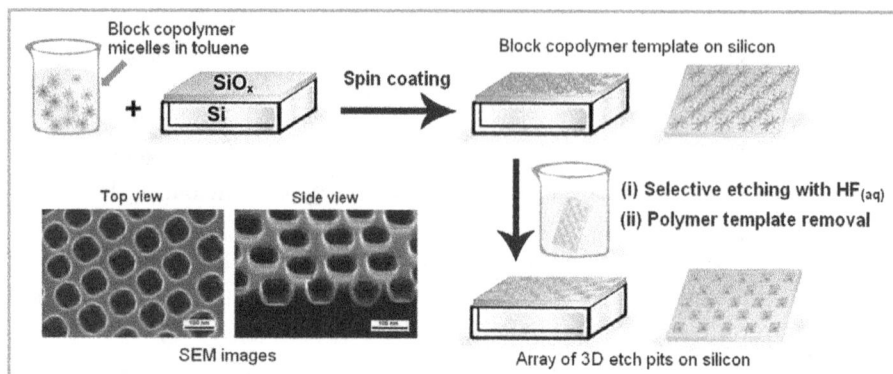

Figure 2.24: Schematic representation of the templated etching techniques for synthesis of a well-defined 3D etches pit array on silicon using a block copolymer as a template and an aqueous solution of hydrofluoric acid (HF) as an etchant. SEM images displaying the etch pit arrays on silicon (adapted from ref. [75]).

2.8.7 Thermal decomposition method

The decomposition temperature, or thermolysis, is the point at which heat is applied to a molecule and breaks down its chemical bonds into individual atoms. This is a straightforward one-step method for synthesizing nanostructures with controlled 3D porosity, as demonstrated by Yu et al. [76] in their creation of nanoporous cadmium oxide (CdO) from cadmium carbonate ($CdCO_3$) microcrystals, as shown in Figure 2.25. This mechanism uses $CdCO_3$ microcrystals as a substrate, which is then covered in CdO starting from their active edges and corners. The growth of the CdO furthers outward to inward across the surface of the microcrystal. Then the material is subjected to heat at 500 °C for 30 min, which initiates the decomposition process. This process involves the release of CO_2 molecules starting from the more exposed surfaces of the sharp edges and corners and gradually proceeding to the flat surfaces of the crystals, ultimately resulting in a core-shell microstructure. The presence of a thick outer layer of CdO effectively obstructs the release of CO_2 molecules from the internal $CdCO_3$ core. As the decomposition of $CdCO_3$ progresses, CO_2 gas accumulates within the microstructure until the internal pressure becomes high enough to facilitate the formation of tiny pores. These pores then grow into continuous channels as the decomposition of $CdCO_3$ is completed. Interestingly, despite significant interior structure changes, the starting microcrystals' overall morphology is largely preserved.

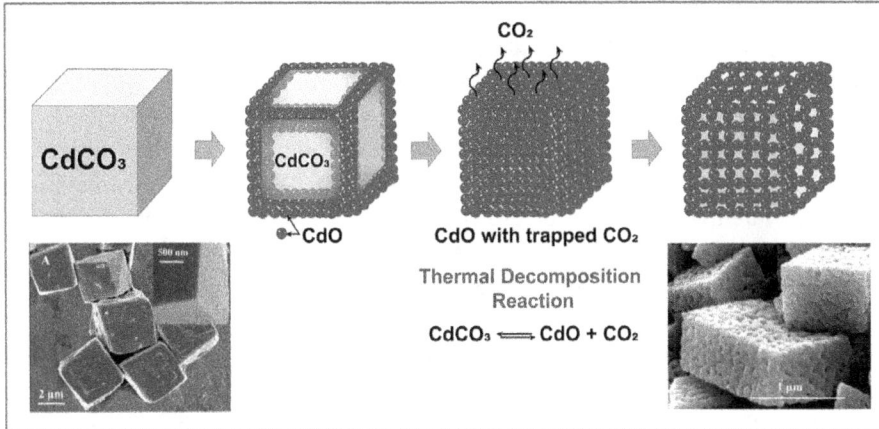

Figure 2.25: Schematic representation of the formation of nanoporous cadmium oxide architectures through one-step thermal decomposition of high-quality cadmium carbonate microcrystals (reproduced from ref. [76]).

2.8.8 Selective dealloying method

Selective dealloying, also known as demetallification, is a surface technique used to selectively remove specific metals from a solution of multiple metals (a metal alloy). This process involves immersing the metal alloy in a chemical solution that is capable of dissolving the unwanted metals elements while leaving the desired metal structure intact. It commonly produces nanoporous metals by utilizing chemical or electrochemical reagents to dissolve the most active metal in the alloy. Selective dealloying involves the removal of the non noble metal, resulting in atoms of the noble metal with three-dimensional (3D) network of pores and channels or ligaments. This process generates intricate structures and patterns, improving the corrosion resistance, wear resistance, and biocompatibility of metal surfaces. Common dealloying processes include (i) decarburization, (ii) decobaltification, (iii) denickelification, (iv) dezincification, and (v) graphite corrosion.

(i) **Decarburization**: Selective carbon loss refers to the removal of carbon atoms from the surface layer of an alloy containing carbon, resulting from a chemical substances with one or more substances in a medium that comes into contact with the surface.

(ii) **Decobaltification**: Selective cobalt removal (or leaching) is the process of extracting cobalt atoms from a cobalt-based alloy, such as satellite or cemented carbides.

(iii) **Denickelification**: Nickel is selectively removed (or leached) from nickel-containing alloys, most commonly observed in copper-nickel alloys after prolonged service in fresh water.

(iv) **Dezincification**: Selective zinc removal (or leaching) occurs in zinc-containing alloys, especially in copper-zinc alloys with less than 85% copper content. This process takes place over time in water with dissolved oxygen, where zinc atoms are efficiently extracted and removed while the other alloy components remain intact.
(v) **Graphitic corrosion**: It is a process of deterioration in grey cast iron where the metallic components undergo selective leaching or conversion to corrosion products, while the graphite structure remains intact. This term is also used to describe the formation of graphite in iron or steel, which usually results from the decomposition of iron carbide at elevated temperatures.

A common example of nanofabrication using dealloying technology is forming nanoporous gold (NPG) by removing silver (Ag) from an $Au_{35}Ag_{65}$ alloy through nitric acid treatment. As shown in Figure 2.26, the size of the nanopores in the NPG increases with longer dealloying times or higher temperatures [77].

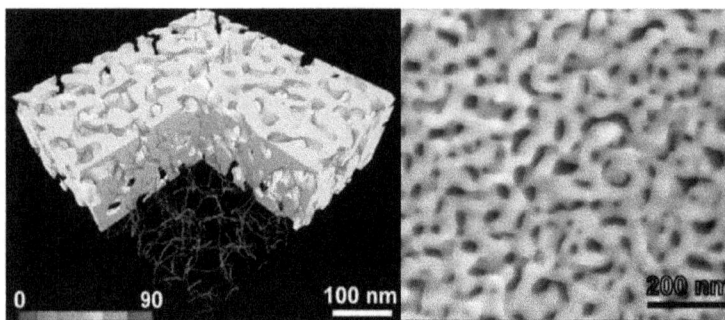

Figure 2.26: Selective dealloying method, electron tomography image of NPG prepared by selective dissolution of silver from the $Au_{35}Ag_{65}$ alloy in nitric acid under free corrosion conditions (adapted from ref. [75]).

2.9 Bottom-up approach

The bottom-up approach in producing nanostructures typically involves the assembly of building blocks, such as atoms or molecules, into ordered arrays due to attractive forces. This approach has the potential to create functional, multicomponent devices through self-assembly without wasting material or requiring the removal of parts. The assembly of these building blocks can be manipulated through physical aggregations, chemical reactions, or the use of templates, allowing controlled chemical reactions to guide the self-assembly process and create nanostructures like nanotubes, nanoribbons, and quantum dots. The bottom-up approach in nanofabrication has the potential to assemble nanostructures that may be difficult to achieve using top-down methods. However, a major challenge in this approach is ensuring predefined structures with precise shapes and

sizes [27]. Despite their potential, bottom-up nanofabrication methods have some limitations, such as the use of harmful reagents and the challenges in achieving precise control
over NP size. Biological routes for nanofabrication also have their own limitations, including being restricted to specific chemical compositions.

2.9.1 Sol-gel deposition in liquid phase

The sol-gel process refers to a wet chemical technique utilized for creating a variety of
inorganic materials, such as metal oxides and ceramics. This process involves the conversion of precursor molecules from monomers to oligomeric suspensions (sol) and subsequently to a polymeric network of gel (gel-like) structures, which form a 3D network
[27]. The sol-gel process involves dissolving molecular precursors, typically metal alkoxides, in water or alcohol. The mixture undergoes gel formation through heating or stirring, known as hydrolysis/alcoholysis. After gelation, the resulting materials undergo
further processing including drying (if alcohol is used) and powdering. Finally, calcination transforms the gels into aerogels, xerogels, or cryogels, depending on the processing
conditions. Aerogels are porous materials with a dense, 3-D structure characterized by
their high specific surface area (SSA) and porosity. Xerogels are also porous structures,
classified into two types based on their skeletal structure: foam and network. Foam xerogels are formed in water-based dispersions using freeze-thawing or air-bubbling techniques. This results in a high porosity of greater than 85% but with a relatively low SSA
that ranges between 4 and 30 m^2/g.

In contrast to foam xerogels, a network-type xerogel possesses a high SSA and low
porosity, approximately 200 m^2/g and 60%, respectively [78]. On the other hand, cryogels
are macroporous cellular structures of hydrogels featuring pores in the size range of
75–300 μm range [79]. The different types of gel are determined by the method used to
remove the solvent (Figure 2.27). Three common methods are utilized: (a) supercritical
drying (SCD), (b) thermal drying, and (c) freeze-drying. SCD is the most used method for
transforming gels into aerogels, and it involves using temperature and pressure to remove the liquid solvent without causing surface tension and capillary stress. Thermal
drying is drying a wet gel at temperatures ranging from 35 to 100 °C, forming xerogels
[79]. Due to its numerous advantages, the sol-gel method has gained significant industrial
recognition for NM synthesis. Key benefits include the ability to produce high-quality
NMs consistent in size, simultaneous manufacturing of multiple types of NMs, fabrication
of homogenous composites with exceptionally high purity (99.99% purity), capability to
design and control chemical composition, production of amorphous materials in thin
layers, creation of materials with modified physical properties like thermal expansion coefficient, low UV absorption, and high optical transparency, high chemical reactivity of
precursors in solution phase, and the use of lower temperatures (70–320 °C) during the
synthesis compared to conventional methods ($T = 1{,}400$–$3{,}600$ °C).

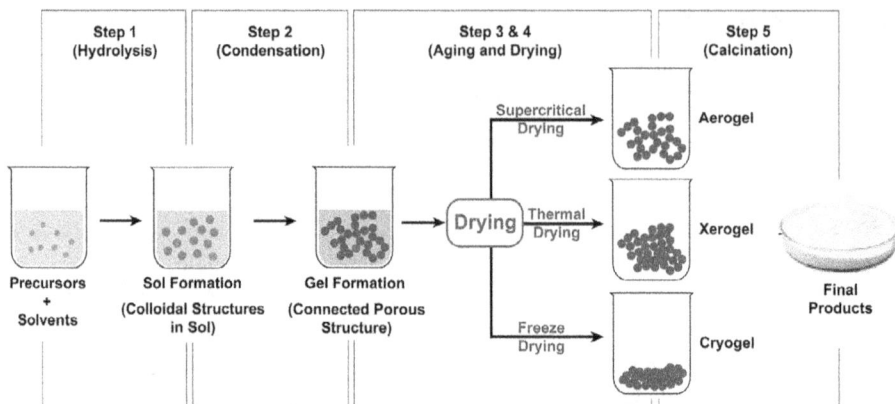

Figure 2.27: Schematic representation of different stages of the sol-gel method: from precursor to final products.

As demonstrated by Barros et al. [79] the creation of hybrid cryogels involved dissolving biopolymers (Gelzan CM, gelatin, and carboxymethylcellulose) in water at a concentration of 3 wt.% to form solution I, while lignin was dissolved in a 1 M sodium hydroxide solution (solution II). The two solutions were mixed in a ratio of 1:4 (w/w) to form the hybrid hydrogels, as shown in Figure 2.28. Calcium carbonate was added at a ratio of 1.823 mmol^{-1} per sodium alginate (on a dry basis). The mixture, consisting of the biopolymer, lignin, and calcium carbonate solutions, was homogenized for a minute using an Ultra-turrax and transferred to a 48-well plate. The suspension was transformed into hydrogels by exposure to supercritical carbon dioxide. Once prepared, the hydrogels were demolded and exposed to freezing under two methods: slow freezing at −80 °C and 3.5 Pa for 24 h, and rapid freezing in liquid nitrogen.

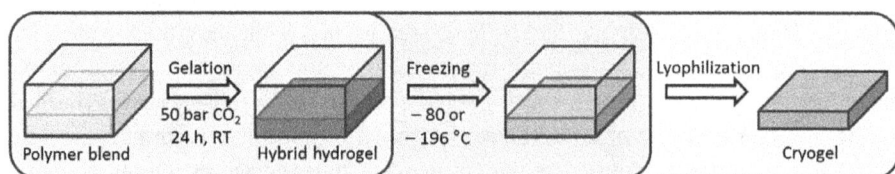

Figure 2.28: Schematic representation of the steps for the synthesizing of cryogens (adapted from ref. [79]).

2.9.2 Atomic layer deposition (ALD)

Atomic layer deposition (ALD) is an advanced gas-phase deposition technique that enables the formation of high-quality, uniformly thin films on substrates. This process functions by repeatedly exposing the substrate to different reactant precursors sequentially, as illustrated in Figure 2.29. ALD's layer-by-layer approach facilitates precise film thickness and composition control, thereby producing thin films with desired properties. The simplest ALD process utilizes a binary (AB) precursor system, where the first precursor (A) reacts with all the available surface sites on the substrate to form a single monolayer and self-limited. Following the reaction of the first precursor (A) with the substrate surface, the remaining precursor A is evacuated from the reaction chamber through purging gas. Thereafter, the second precursor (B) is introduced into the chamber to react with the monolayer of grafted A, once again in a self-limiting manner, which completes one cycle of the ALD process as depicted in Figure 2.29. Common precursor A compounds used in the deposition of metal oxides include organometallic compounds, such as alkoxides and alkyls, with water being the most frequently used precursor B. However, other reagents can also be employed [27].

For example, ammonia (NH_3) can be utilized as a precursor and catalyst in the process of SiO_2 deposition, as silicon precursors like silicon tetrachloride ($SiCl_4$) and tetraethoxysilane (TEOS) typically exhibit low reaction rates within the temperature range achievable with most ALD apparatuses (30–250 °C). On the other hand, trimethylaluminum (TMA) reacts rapidly and effectively with water (H_2O) even at room temperature, making TMA-H_2O a widely used precursor pair for the deposition of Al_2O_3.

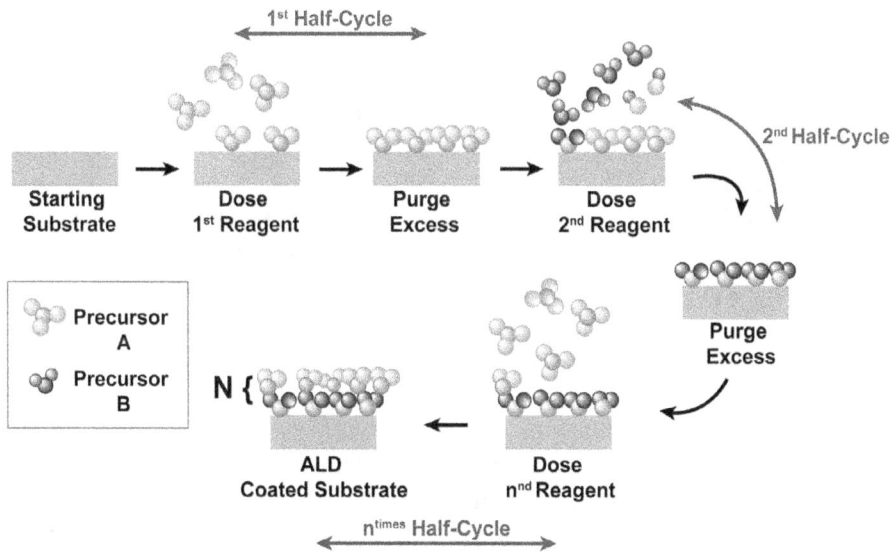

Figure 2.29: Schematic illustration of a typical ALD cycle consisting of sequential precursor and co-reactant doses are separated by purge or pump steps, leading to self-limiting film growth. N indicates the metal atom, which can form N layers.

The reactions between the precursors and the substrate in an ALD process are surface-limited, meaning that each pulse reacts with the available surface sites until the

reaction self-limits. This ensures that all active sites on the substrate are used without excess reaction, and the process stops. The nature of the interaction between the precursor molecules and the substrate surface determines the complete ALD cycle. The ALD cycle can be repeated multiple times to build up the desired thickness of the thin film. ALD processes are often carried out at lower temperatures, which is advantageous when working with fragile substrates. Furthermore, certain precursor materials that have thermal instability can still be used in ALD processes as long as their decomposition rate is slow. ALD is capable of depositing a wide range of materials including oxides, metals, sulfides, and fluorides, and the coatings can possess a broad range of properties depending on the intended application. ALD coatings are frequently used for providing ultra-thin nanolayers with exceptional precision on various substrates, even micron-to-micron particles. The nanolayers produced through the ALD process exhibit several required features including the following:
- Atomic layer accuracy and precision in thickness control
- Conformal coverage even on high-aspect ratio and high-surface-area materials
- Very low defect density and high uniform coatings
- Ability to create complex and multilayered films such as nanolaminates, nano-alloys, and precise doping
- Versatility in producing various insulating, conducting, and semiconducting films
- Scalability for industrial-scale applications.

Gould et al. [80] synthesized nickel nanoparticles on Al_2O_3 support by using the ALD method in a fluidized bed reactor. The process first involved reacting bis(cyclopentadienyl)Ni with an N_2 carrier gas on the alumina surface and then reacting the organometallic molecules bound to the surface with a 20% H_2 (balance Ar) mixture. By adjusting the number of ALD cycles, we can control the amount of Ni deposited between 4.7 and 16.8 wt.%, therefore, the size of the Ni particles between 2.4 and 3.3 nm on average. Wang et al. [81] aimed to enhance the durability of nickel zeolite catalysts by diffusing the Pt-metal in the KL zeolite, it is a zeolite L (LTL) using potassium (K) as the cationic species, with the ALD method. The support was dispersed in ethanol (200 mL) and dropped onto a square quartz plate, which was then dried at room temperature. Next, $Me_3(MeCp)Pt$ and O_3 were sequentially pulsed into the ALD reactor which contained KL zeolite. For Me_3(MeCp)Pt, the pulse time was 0.5 s, and the exposure time was 60 s, while the purge time was 100 s. Six ALD cycles were performed to obtain precisely deposited Pt on the surface of KL zeolite. To achieve sufficient vapor pressure, the stainless steel cylinder containing the Pt precursor was maintained at 65 °C, and the reactor temperature was kept at 220 °C. The average diameter of the Pt crystals in the (100) direction and the length in the (001) direction have been determined in the obtained sample. In other effort Li et al. [82] sought to employ the ALD method to synthesize zeolite-based catalysts for the purpose of designing and controlling the structure of active components. However, the complex nature of zeolite channels and defects poses significant challenges for ALD technology when it comes to the design, regulation, and application of large-scale zeolite-based catalysts.

2.9.3 Chemical vapor deposition (CVD)

In CVD, a solid material or volatile precursors are deposited on an exposed substrate. This is achieved through the reaction and/or decomposition of the precursors on the substrate's surface, forming a thin film deposit. Numerous factors, such as the chosen substrate material, system setup, reactor configuration, gas feedstock, gas ratios, reactor pressure, gas partial pressures, reaction temperature, growth time, substrate temperature, and the composition of the reaction gas mixture, total pressure gas flows, and other materials with a wide range of physical, tribological, and chemical properties, can be produced as illustrated in Figure 2.30.

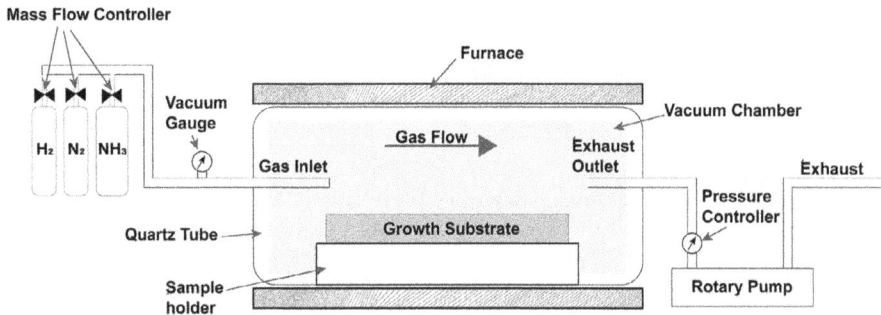

Figure 2.30: Schematic representation of chemical vapor deposition (CVD).

During the CVD process, several main steps occur:
1. Precursor generation: The precursor is converted into active gaseous reactant species.
2. Transport: The precursor, in gaseous form, is delivered into the reaction chamber.
3. Adsorption: The precursor gas molecules are adsorbed onto the hot surface.
4. Decomposition: The precursor decomposes and releases the desired atoms required for film formation while generating organic waste.
5. Migration: The atoms move to strong binding sites.
6. Nucleation: This leads to the growth of the thin film, the desorption of unwanted side products, and the removal of unwanted products.

With its outstanding throwing power, the rotary CVD method produces coatings with uniform thickness and properties, exhibiting a minimal porosity level. This process ensures high-quality CVD output and enables localized or selective deposition, especially on patterned substrates. The CVD process can be categorized into seven distinct types based on the following factors: temperature, pressure, wall/substrate, precursor nature, depositing time, gas flow state, and activation/power source [83], as depicted in Figure 2.31.

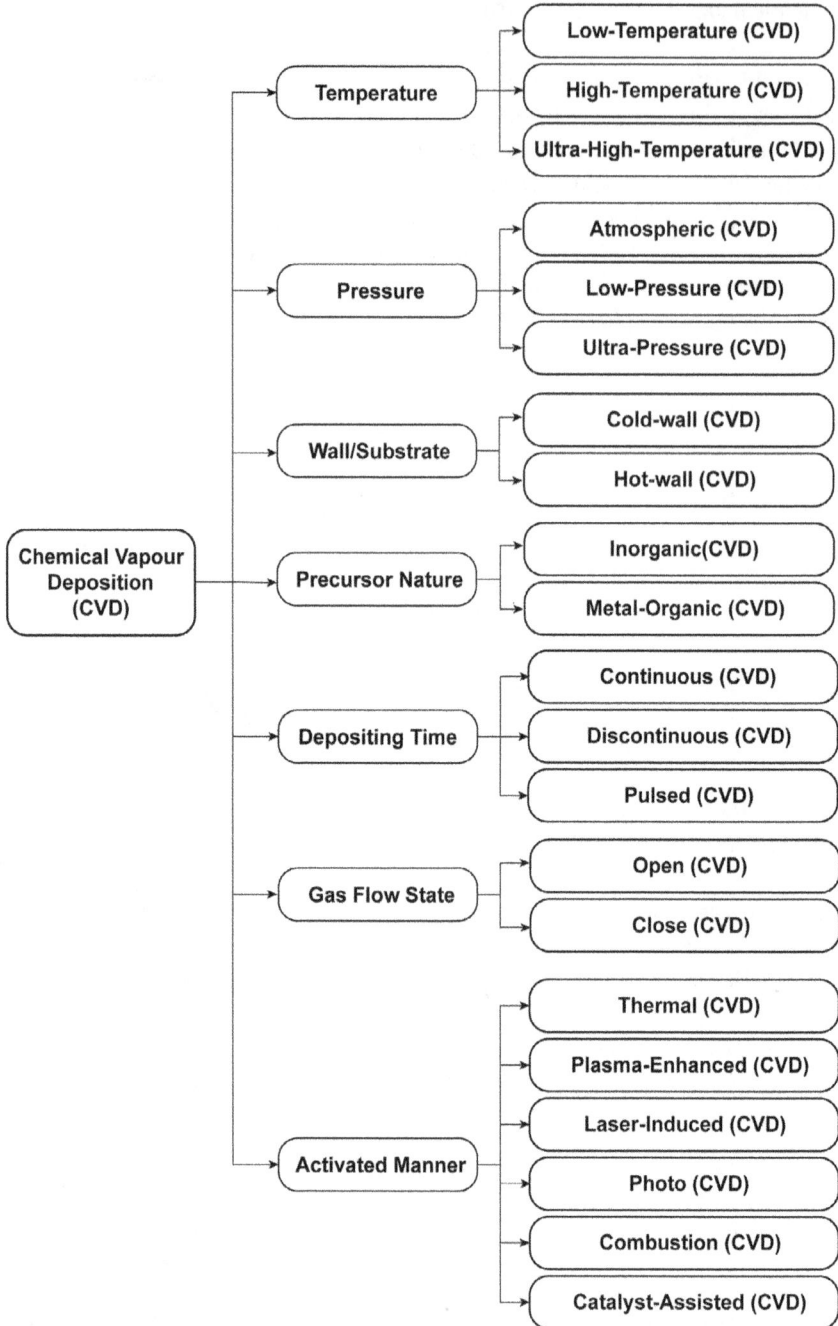

Figure 2.31: Schematic representation of the classification of CVD [83].

The adaptability and wide-ranging capabilities of CVD methods are undeniably impressive, as they are used in diverse fields. They are utilized in the formation of thin films (such as dielectrics, conductive oxides, conductors, passivation layers, tribological and corrosion-resistant coatings, epitaxial layers for microelectronics, and heat-resistant barriers), synthesizing high-temperature materials (including tungsten, ceramics, and others), as well as the creation of advanced 2D nanosheets such as graphene, h-BN nanosheets, borophenes, and metal carbides. The versatility and adaptability of CVD processes make them an intriguing study area for further exploration and innovation [27].

This process, as described in the study by Liu et al. [84] was a meticulous procedure involving carefully controlling the coupled mass transport and chemical reaction of carbon precursors in the synthesis of zeolite-templated carbon (ZTC) using the CVD method. A vertical fixed-bed reactor with a size of 5.5 cm was employed in the process, and zeolite-13X was placed within the reactor and heated to 600 °C under a nitrogen atmosphere at a flow rate of 600 cm^3/min. To obtain ZCP samples, the nitrogen flow was subsequently changed to a mixture of C_2H_2 and nitrogen (with a volume concentration of 2% of C_2H_2 in the C_2H_2/N_2 mixture), following an isothermal phase in which the furnace temperature was maintained at 600 °C for 30 min. This process was continued for another 5 h. After that, the furnace temperature was increased to 900 °C under an N_2 gas flow to achieve a denser carbon framework in the ZPC sample. The final stage involved removing the zeolite template from the ZCP sample using an aqueous solution of HF and HCl (with 1.1 and 0.8 M concentrations, respectively). The template was removed through several washing and filtration steps using DI water, which resulted in the formation of ZTC. In this process, the CVD apparatus was operated at a heating rate of 5 °C/min and a gas flow rate of 600 cm^3/min. In the process described by Ali et al. [85] a composite material consisting of nNi/NiO-multilayered graphene was synthesized through the CVD method. First, Ni-NPs were placed in a ceramic boat and then in a double-zone tubular CVD reactor. The CVD chamber was initially flushed with Ar gas at a 100 cm/min flow rate for 10 min to purge the air in the chamber. Following this, the chamber was filled with Ar gas at a flow rate of 300 sccm until the chamber pressure reached ~122.6 kPa. Next, the heating zone in the chamber was gradually heated to 850 °C, increasing the temperature at a rate of 10 °C/min. Once the required temperature was reached, the ceramic boat containing the catalyst particles was moved from its initial position toward the center of the heating zone. Next, acetylene gas was introduced into the chamber at a flow rate of 150 cm, and it was allowed to undergo pyrolysis for 10 min. Throughout the reaction, the chamber pressure was measured at approximately 147 kPa. Once the reaction was complete, the acetylene gas flow was interrupted, and the Ar flow rate was decreased to 100 cm and the chamber was allowed to cool to room temperature under an Ar flow. In case of Oyama et al. [86], silica-based membranes were synthesized using the CVD of TEOS on Y-alumina overlayer. This process involved the creation of three layers: a commercial α-alumina support with a nominal pore size of 60 nm, an intermediate layer of Y-alumina deposited by sequential dip-coating and calcining, and a topmost silica layer deposited by CVD. The topmost silica layers were relatively thin (20–30 nm) and were formed by the thermal decomposition of the silica precursor compound. The process began with optimizing the intermediate layer by utilizing conventional silica precursor TEOS. A 6-mm diameter silica support tube was co-axially fixed inside a stainless reactor using machined Swagelok fittings with Teflon ferrules. The assembly was placed inside an electric furnace and gradually heated to 650 °C at a rate of 1.5 °C/min. The bubbler temperature for delivering TEOS was kept at 25 °C, resulting in a TEOS vapor pressure of 250 Pa.

2.9.4 Laser pyrolysis synthesis

Laser pyrolysis can be regarded as an exclusive synthesis technique that involves the interplay between laser photons, a gaseous reactant species, and a sanitizer agent. This method, often classified as a vapor-stage synthesis technique, is particularly proficient in generating NMs. The fundamental principle underlying laser pyrolysis lies in the resonance between the emission line of a CO_2 laser and the infrared absorption line of a gas- or vapor-based precursor molecule. In cases where the precursor lacks an absorption band at the required wavelength, the method relies on a separate agent to play the role of a sensitizer. Typically, a sensitizer like ethylene (C_2H_4) absorbs the energy from the laser radiation and subsequently transfers it to the precursor through collisions. The overall structure of the synthesis apparatus is based on a crossflow configuration within a chamber, wherein reactants are introduced into the chamber through a carrier gas (such as Argon). Once inside, the gaseous-phase precursors encounter the laser beam, with the high-power laser beam (e.g., 2,400 W) generates elevated localized temperatures, which trigger the nucleation and growth of NMs. The final step in the process involves the reactant flow intersecting with the laser beam, which is continuously focused as a wave of CO_2 laser radiation. The resulting nanoparticles are collected by a catcher equipped with a filter [27], as illustrated in Figure 2.32. The collected material is then passed through porous filtering barriers before being washed with acetone to remove any side products of the synthesis, which are believed to consist of polycyclic aromatic hydrocarbons [87].

Laser pyrolysis boasts numerous advantages over conventional synthesis methods, mainly attributed to the following reasons: (i) it is a one-step process, (ii) it yields high purity final products, (iii) it offers high yield and reproducibility, (iv) it requires no additional steps for sample preparation, and (v) it allows easy control of experimental parameters. The versatility of producing nanostructures through laser pyrolysis depends largely on the appropriate choice of precursor and the optimal tuning of experiment parameters. Therefore, selecting effective precursors is crucial for achieving the desired NPs [88]. As an example, D'Amato et al. [89] successfully generated SiC, TiO_2, and SiO_2 nanoparticles using laser pyrolysis, while Karpiel et al. [90] synthesized Cu_xO_y/TiO_2 under either He or Ar gas. The main principle behind this reaction is the orthogonal interaction between an infrared CO_2 laser in continuous mode and a

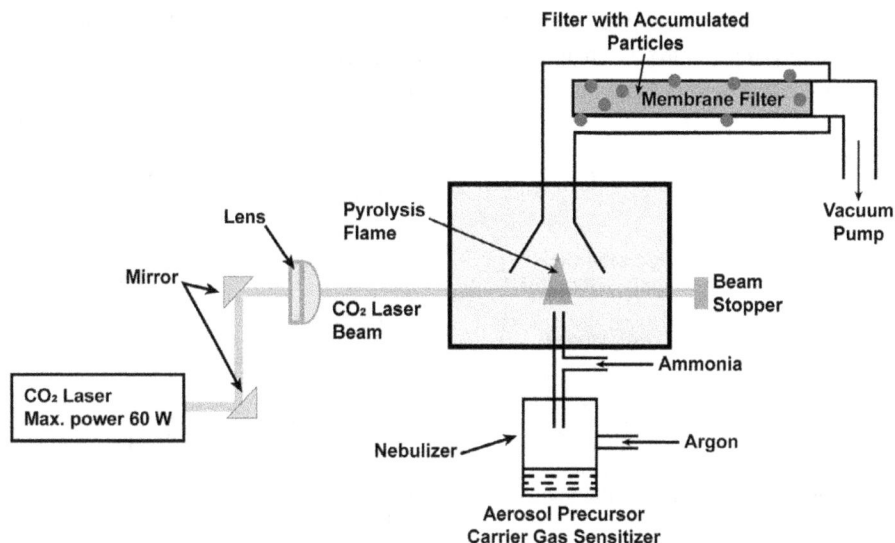

Figure 2.32: A schematic representation of the experimental setup for CO_2 laser pyrolysis.

gaseous or liquid mixture of reagents. Focusing of the laser beam is achieved by using a cylindrical lens, resulting in a horizontal beam with a diameter of approximately 30 mm. For the synthesis of samples (TiO_2), the mixture comprises liquid titanium (IV) isopropoxide (TTIP) dissolved in o-xylene/ethyl acetate. Introducing a sensitizer gas (C_2H_4) into the carrier gas (He or Ar) enhances the laser absorption of the precursors. The CO_2 laser's power was maintained at 670 W, and the pressure in the reactor was kept at a constant value of 740 Torr. Particles of the synthesized material were gathered on filters situated downstream of the reaction zone.

2.9.5 Hydrothermal synthesis

Hydrothermal synthesis is a chemical technique for generating materials at temperatures above 100 °C and at least 100 bar pressures using either hydrothermal or solvothermal conditions. This method is influenced by geological processes that occur deep within the earth's crust. Under these elevated temperature and pressure conditions, the process produces unique crystal structures and porous, inaccessible phases that could not be formed through conventional methods.

This hydrothermal technique is especially effective for creating materials that need specific growth conditions. It is a powerful tool for forming inorganic porous materials with controlled structures or creating mesoporous materials with improved properties. Additionally, this technique allows the synthesis of NMs that normally would be unstable at elevated temperatures, and it minimizes the loss of materials, especially for those with high vapor pressures. The method can produce nanoparticles with unique morphologies, structures, and porous features such as metal oxide, zeolite, zeolite-like, and perovskite, as depicted in Figure 2.33.

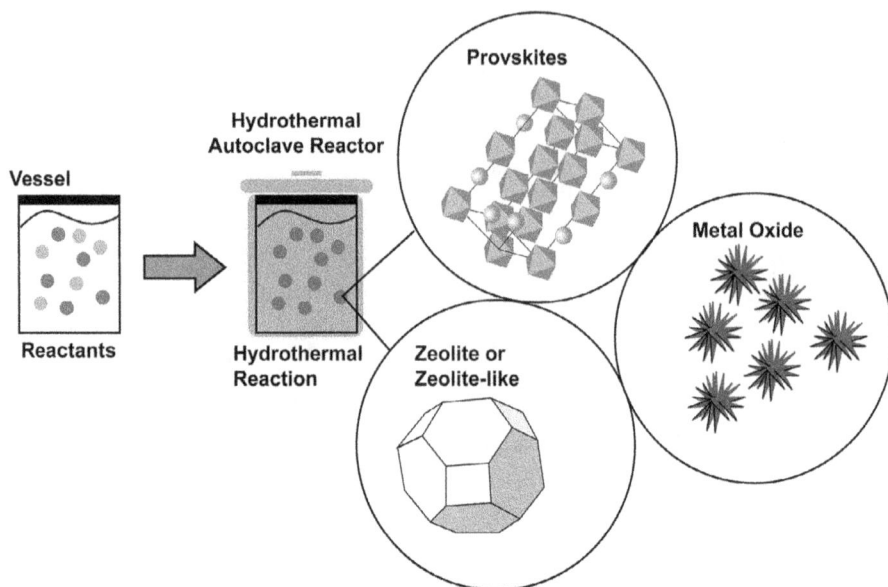

Figure 2.33: A schematic representation of the steps in hydrothermal synthesis of nanomaterials.

For instance, Zhang et al. [91] developed Ag-nano cellulose nanocomposites using nano cellulose in an aqueous solution, where the nano cellulose acted as a reducing agent and stabilizer. They first dissolved microcrystalline cellulose, (2,2,6,6-Tetramethylpiperidin-1-yl)oxyl (TEMPO), and NaBr in pure water and heated the solution in a water bath at 40 °C while continuously stirring. Next, the sodium chlorite solution was added, and the pH of the solution was adjusted to 10 with the addition of NaOH solution. The oxidation product was collected by centrifuging the solution at 10,000 rpm for 10 min. Then, the obtained precipitate was dissolved with the help of sonication. The solution was centrifuged at 3,000 rpm to remove any remaining large impurities. After that, the nano cellulose solution and various volumes of $AgNO_3$ were mixed together. The final mixture was transferred to a 50-mL teflon-lined autoclave and heated at 140 °C for 4 h. The material obtained is a yellowish hue. Moreover, hydrothermal techniques can also synthesize metal oxides, as demonstrated by Basahel et al. [92]. Their study involved the synthesis of nanosized ZrO_2 powders featuring different tetragonal and cubic structures through hydrothermal methods. The tetragonal structure was created by using zirconium oxychloride, which was dissolved in an ammonia solution and combined with DI water followed by stirring at room temperature for a duration of 4 h. Next, the obtained precipitate was separated through centrifugation, washed with water and ethanol, and then transferred into a Teflon-lined autoclave at a temperature of 100 °C for a period of 12 h. The cubic ZrO_2 powder was synthesized by dissolving zirconium isopropoxide in ethanol, followed by transferring the solution into a Teflon-lined vessel and keeping it in a desiccator with water at the bottom for 12 h, followed by added NaOH solution to the mixture. Then, the mixture solution was transformed into an autoclave at 180 °C for 18 h.

Zeolite and zeolite-like materials can also be synthesized through hydrothermal methods. For instance, the analcime zeolite was synthesized via hydrothermal processing. Various aluminum precursors, such as aluminum nitrate, aluminum sulfate anhydrous, aluminum isopropoxide, and sodium aluminate, were combined with Si precursors and agitated vigorously. Subsequently, an aqueous solution of NaOH was added, and the mixture was stirred for a period of 4 h until a homogeneous solution was achieved. The resulting mixture was transferred into an autoclave for the crystallization process to take place under hydrothermal conditions at a temperature of 190 °C for 72 h. Once the hydrothermal treatment was complete, the produced material was washed with distilled water using a vacuum filtration process until all residual NaOH was removed. Finally, the obtained cake was dried in an oven at 70 °C for 24 h [93]. For the synthesis of SAPO-34 (zeolite-like material), pseudo-boehmite was dissolved in DI water and then stirred at room temperature for 10 min. Subsequently, phosphoric acid was added to the solution, and the mixture was further agitated for 1 h to form a gel. Subsequently, the citric acid solution was added to the mixture and stirred vigorously for 10 min at room temperature. Then, colloidal silica and triethylamine solution were gradually added to the final gel and stirred again for 4 h. The completed gel was then transferred to an autoclave and subjected to further processing at 180 °C for varying periods [94]. In addition, perovskite oxides such as $LaFeO_3$ can be synthesized using hydrothermal and solvothermal processes. Gomez-Cuaspud et al. [95] successfully synthesized $LaFeO_3$ oxide through hydrothermal processing. First dissolved nitrates of lanthanum nitrate hexahydrate, $(La(NO_3)_3.6H_2O)$, and ferric nitrate nonahydrate, $(Fe(NO_3)_3.9H_2O)$, in a 1.0 mol/L concentration. Subsequently, the nitrate mixture was combined in a Teflon-lined steel container and potassium hydroxide (KOH) to create an alkaline medium. The hydrothermal reaction was then performed at a temperature of 280 °C for a duration of 4 days, obtaining a black crystalline solid material.

2.10 Biological (green) synthesis

Green synthesis, also known as biological synthesis, utilizes reducing agents derived from biomaterial extracts to reduce metal species in aqueous solutions. Such biomaterials can be sourced from plants, including leaves, roots, flowers, seeds, stems, bark, and fruits, as well as various microorganisms like bacteria, fungi, yeast, and algae. Green synthesis of NMs has become a high-priority area in nanotechnology research, as it aims to minimize the production of harmful byproducts, emphasizing the importance of easy and non-toxic production techniques. This method stands out for its eco-friendly, pure, cost-effective, and versatile nature. Silver and gold NM synthesis often perform biosynthesis methodology as a pivotal advancement in nanotechnology.

2.11 References

[1]   G. Palmisano, S. Al Jitan and C. Garlisi. *Heterogeneous Catalysis: Fundamentals, Engineering and Characterizations*, Elsevier, 2022.

[2]   P. Unnikrishnan and D. Srinivas, Heterogeneous catalysis: in *Industrial Catalytic Processes for Fine and Specialty Chemicals*, eds. S. S. Joshi, and V. V. Ranade, Elsevier, Amsterdam, 2016, pp. 41–111.

[3]   G. M. Barrow. *Physical Chemistry*, McGraw-Hill Companies, INC, 6th edn., 1996.

[4]   Y. Tian and J. Wu. Differential heat of adsorption and isosteres, *Langmuir*, 2017, **33**, 996–1003.

[5]   I. Men'shchikov, A. Shkolin, E. Khozina and A. Fomkin. Thermodynamics of adsorbed methane storage systems based on peat-derived activated carbons, *Nanomaterials*, 2020, **10**, 1379.

[6]   Y. Changtao, L. Shuyuan, W. Hailong, Y. Fei and X. Xinyi. Pore structure characteristics and methane adsorption and desorption properties of marine shale in Sichuan Province, China, *RSC Adv.*, 2018, **8**, 6436–6443.

[7]   M. Batzill and U. Diebold. The surface and materials science of tin oxide, *Prog. Surf. Sci.*, 2005, **79**, 47–154.

[8]   K. A. Rahman, W. S. Loh and K. C. Ng. Heat of adsorption and adsorbed phase specific heat capacity of methane/activated carbon system, *Procedia Eng.*, 2013, **56**, 118–125.

[9]   A. A. Shaikh. *Heterogeneous Catalysis: Solid Catalysts, Kinetics, Transport Effects, Catalytic Reactors*, Walter de Gruyter GmbH, 2nd edn., 2023.

[10]   K. S. W. Sing. Reporting physisorption data for gas/solid systems with special reference to the determination of surface area and porosity (Recommendations 1984), *Pure & Appl. Chem.*, 1985, **57**, 603–619.

[11]   S. Yurdakal, C. Garlisi, L. Özcan, M. Bellardita and G. Palmisano, eds. G. Marcì and L. B. T.-H. P. Palmisano, Elsevier, 2019, pp. 87–152.

[12]   S. Brunauer, L. S. Deming, W. E. Deming and E. Teller. On a theory of the van der Waals adsorption of gases, *J. Am. Chem. Soc.*, 1940, **62**, 1723–1732.

[13]   L. Hauchhum and P. Mahanta. Carbon dioxide adsorption on zeolites and activated carbon by pressure swing adsorption in a fixed bed, *Int. J. Energy Environ. Eng.*, 2014, **5**, 349–356.

[14]   M. Rahimnejad, S. K. Hassaninejad-Darzi and S. M. Pourali. Preparation of template-free sodalite nanozeolite–chitosan-modified carbon paste electrode for electrocatalytic oxidation of ethanol, *J. Iran. Chem. Soc.*, 2015, **12**, 413–425.

[15]   S. Maiti, A. Pramanik, U. Manju and S. Mahanty. Reversible lithium storage in manganese 1,3,5-benzenetricarboxylate metal–organic framework with high capacity and rate performance, *ACS Appl. Mater. Interfaces*, 2015, **7**, 16357–16363.

[16]   L. Zhou, X. Liu, J. Li, Y. Sun and Y. Zhou. Sorption/desorption equilibrium of methane in silica gel with pre-adsorption of water, *Colloids Surfaces A Physicochem. Eng. Asp.*, 2006, **273**, 117–120.

[17]   Z. Li, K. Wang, J. Song, Q. Xu and N. Kobayashi. Preparation of activated carbons from polycarbonate with chemical activation using response surface methodology, *J. Mater. Cycles Waste Manag.*, 2014, **16**, 359–366.

[18]   J. M. Gay, J. Suzanne and J. P. Coulomb. Wetting, surface melting, and freezing of thin films of methane adsorbed on MgO(100), *Phys. Rev. B*, 1990, **41**, 11346–11351.

[19] P. Llewellyn, E. Bloch and S. Bourrelly, Surface area/ porosity, adsorption, diffusion: in *Characterization of Solid Materials and Heterogeneous Catalysis*, eds. M. Che and J. Védrine, Wiley-VCH Verlag GmbH & Co, Weinheim, 2015, pp. 853–879.

[20] S. Pan, M. Zha, C. Gao, J. Qu and X. Ding. Pore structure and fractal characteristics of organic-rich lacustrine shales of the Kongdian formation, Cangdong Sag, Bohai Bay Basin, *Front. Earth Sci.*, 2021, **9**, 760583.

[21] J. M. Thomas and W. J. Thomas. *Principles and practice of heterogeneous catalysis*, Wiley-VCH Verlag GmbH & Co, 2nd edn, 2015.

[22] J. W. in *Niemantsverdriet*, Appendix: Metal surfaces and chemisorption: *in Spectroscopy in Catalysis: An Introduction*, ed. J. W. Niemantsverdriet, John Wiley & Sons, Weinheim, 2007, pp. 297–320.

[23] T. Sheng, Y.-X. Jiang, N. Tian, Z.-Y. Zhou and S.-G. Sun, Nanocrystal catalysts of high-energy surface and activity, in *Morphological, Compositional, and Shape Control of Materials for Catalysis*, eds. P. Fornasiero, M. B. T.-S. In S. S. and C. Cargnello, Elsevier, Amsterdam, 2017, vol. 177, 439–475.

[24] B. C. Gates. *Catalytic Chemistry*, John Wiley & Sons, Inc, 3rd edn., 1991.

[25] C. Daraio and S. Jin, Synthesis and pattering methods for nanostructures useful for biological applications: in *Nanotechnology for Biology and Medicine*, eds. G. A. Silva and V. Parpura Springer, New York, NY, 2012, pp. 27–44.

[26] C. Dhand, N. Dwivedi, X. J. Loh, A. N. Jie Ying, N. K. Verma, R. W. Beuerman, R. Lakshminarayanan and S. Ramakrishna. Methods and strategies for the synthesis of diverse nanoparticles and their applications: A comprehensive overview, *RSC Adv.*, 2015, **5**, 105003–105037.

[27] N. M. Noah. Design and synthesis of nanostructured materials for sensor applications, *J. Nanomater.*, 2020, **2020**, 8855321.

[28] K. K. Anoop, X. Ni, X. Wang, R. Bruzzese and S. Amoruso. Spectrally resolved imaging of ultrashort laser produced plasma, *IEEE Trans. Plasma Sci.*, 2014, **42**, 2698–2699.

[29] A. Bogaerts, Z. Chen, R. Gijbels and A. Vertes. Laser ablation for analytical sampling: What can we learn from modeling?, *Spectrochim. Acta Part B At. Spectrosc.*, 2003, **58**, 1867–1893.

[30] N. M. Bulgakova and A. V. Bulgakov. Pulsed laser ablation of solids: Transition from normal vaporization to phase explosion, *Appl. Phys. A.*, 2001, **73**, 199–208.

[31] A. Nyabadza, É. McCarthy, M. Makhesana, S. Heidarinassab, A. Plouze, M. Vazquez and D. Brabazon. A review of physical, chemical and biological synthesis methods of bimetallic nanoparticles and applications in sensing, water treatment, biomedicine, catalysis and hydrogen storage, *Adv. Colloid Interface Sci.*, 2023, **321**, 103010.

[32] A. Nyabadza, M. Vazquez and D. Brabazon, A review of bimetallic and monometallic nanoparticle synthesis via laser ablation in liquid, *Crystals*, 2023, **13**, 253.

[33] M.-R. Kalus, R. Lanyumba, S. Barcikowski and B. Gökce. Discrimination of ablation, shielding, and interface layer effects on the steady-state formation of persistent bubbles under liquid flow conditions during laser synthesis of colloids, *J. Flow Chem.*, 2021, **11**, 773–792.

[34] R. Streubel, G. Bendt and B. Gökce. Pilot-scale synthesis of metal nanoparticles by high-speed pulsed laser ablation in liquids, *Nanotechnology*, 2016, **27**, 205602.

[35] T. Sasaki, Y. Shimizu and N. Koshizaki. Preparation of metal oxide-based nanomaterials using nanosecond pulsed laser ablation in liquids, *J. Photochem. Photobiol. A Chem.*, 2006, **182**, 335–341.

[36] A. Nyabadza, M. Vázquez and D. Brabazon. Modelling of pulsed laser ablation in liquid via Monte Carlo techniques: The effect of laser parameters and liquid medium on the electron cloud, *Solid State Sci.*, 2022, **133**, 107003.

[37] A. Nyabadza, M. Vázquez, S. Coyle, B. Fitzpatrick and D. Brabazon, Magnesium nanoparticle synthesis from powders via pulsed laser ablation in liquid for nanocolloid production, *Appl. Sci.*, 2021, **11**, 10974.

[38] C. L. Sajti, R. Sattari, B. N. Chichkov and S. Barcikowski. Gram scale synthesis of pure ceramic nanoparticles by laser ablation in liquid, *J. Phys. Chem. C*, 2010, **114**, 2421–2427.

[39] A. K. H Ferman, G. H S. Jaber and W. Mahmood. Fabrication of zinc oxide nanoparticles by pulsed laser ablation in liquid – PLAL and study their physical properties, *J. Phys. Conf. Ser*, 2021, **1900**, 12008.

[40] T. Hesabizadeh, K. Sung, M. Park, S. Foley, A. Paredes, S. Blissett and G. Guisbiers, Synthesis of antibacterial copper oxide nanoparticles by pulsed laser ablation in liquids: potential application against foodborne athogens, *Nanomaterials*, 2023, **13**, 2206.

[41] O. Y. Griaznova, I. B. Belyaev, A. S. Sogomonyan, I. V. Zelepukin, G. V. Tikhonowski, A. A. Popov, A. S. Komlev, P. I. Nikitin, D. A. Gorin, A. V. Kabashin and S. M. Deyev, Laser synthesized core-satellite Fe-Au nanoparticles for multimodal in vivo imaging and in vitro photothermal therapy, *Pharmaceutics*, 2022, **14**, 994.

[42] M. A. Almessiere, Y. Slimani, Y. O. Ibrahim, M. A. Gondal, M. A. Dastageer, I. A. Auwal, A. V. Trukhanov, A. Manikandan and A. Baykal. Morphological, structural, and magnetic characterizations of hard-soft ferrite nanocomposites synthesized via pulsed laser ablation in liquid, *Mater. Sci. Eng. B*, 2021, **273**, 115446.

[43] Z. Deng, T. J. Pisklak and K. J. Balkus. The preparation of partially oriented zeolite thin films via pulsed laser ablation, *MRS Online Proc. Libr.*, 2003, **752**, 22.

[44] M. S. El-Eskandarany, A. Al-Hazza, L. A. Al-Hajji, N. Ali, A. A. Al-Duweesh, M. Banyan and F. Al-Ajmi, Mechanical milling: A superior nanotechnological tool for fabrication of nanocrystalline and nanocomposite materials, *Nanomaterials*, 2021, **11**, 2484.

[45] T. P. Yadav, R. M. Yadav and D. P. Singh. Mechanical milling: A top down approach for the synthesis of nanomaterials and nanocomposites, *Nanosci. Nanotechnol.*, 2012, **2**, 22–48.

[46] S. Dhiman, R. S. Joshi, S. Singh, S. Gill, H. Singh, R. Kumar and V. Kumar. A framework for effective and clean conversion of machining waste into metal powder feedstock for additive manufacturing, *Clean. Eng. Technol.*, 2021, **4**, 100151.

[47] S. M. Almotairy, N. H. Alharthi and H. S. Abdo. Regulating mechanical properties of Al/SiC by utilizing different ball milling speeds, *Crystals*, 2020, **10**, 332.

[48] C. Suryanarayana. Mechanical alloying and milling, *Prog. Mater. Sci.*, 2001, **46**, 1–184.

[49] R. S. Fediuk, R. A. Ibragimov, V. S. Lesovik, A. A. Pak, V. V. Krylov, M. M. Poleschuk, N. Y. Stoyushko and N. A. Gladkova. Processing equipment for grinding of building powders, *IOP Conf. Ser. Mater. Sci. Eng.*, 2018, **327**, 42029.

[50] S.-H. Hong and B.-K. Kim. Effects of lifter bars on the ball motion and aluminum foil milling in tumbler ball mill, *Mater. Lett.*, 2002, **57**, 275–279.

[51] B. Li, L. He, Y. Guo, H. Zhao, J. Shen, W. Lei, J. Xu and H. Shao. High energy ball milling to synthesize transition metal vanadates with boosted lithium storage performance, *Mater. Today Commun.*, 2023, **37**, 107496.

[52] G. Liu, Z. Su, D. He and C. Lai. Wet ball-milling synthesis of high performance sulfur-based composite cathodes: The influences of solvents and ball-milling speed, *Electrochim. Acta*, 2014, **149**, 136–143.

[53] J. Qu, X. He, X. Li, Z. Ai, Y. Li, Q. Zhang and X. Liu. Precursor preparation of Zn–Al layered double hydroxide by ball milling for enhancing adsorption and photocatalytic decoloration of methyl orange, *RSC Adv.*, 2017, **7**, 31466–31474.

[54] A. L. Gobbi and P. A. P. Nascente, D. C. sputtering: in *Encyclopedia of Tribology*, eds. Q. J. Wang and Y.-W. Chung Springer US, Boston, MA, 2013, pp. 699–706.

[55] H. Poelman, K. Eufinger, D. Depla, D. Poelman, R. De Gryse, B. F. Sels and G. B. Marin. Magnetron sputter deposition for catalyst synthesis, *Appl. Catal. A Gen.*, 2007, **325**, 213–219.

[56] V. Orozco-Montes, A. Caillard, P. Brault, W. Chamorro-Coral, J. Bigarre, A. Sauldubois, P. Andreazza, S. Cuynet, S. Baranton and C. Coutanceau. Synthesis of platinum nanoparticles by plasma sputtering onto glycerol: Effect of argon pressure on their physicochemical properties, *J. Phys. Chem. C*, 2021, **125**, 3169–3179.

[57] Y. Xu, G. Li and Y. Xia. Synthesis and characterization of super-hard AlCrTiVZr high-entropy alloy nitride films deposited by HiPIMS, *Appl. Surf. Sci.*, 2020, **523**, 146529.

[58] S. Zhao, C. Liu, J. Yang, W. Zhang, L. He, R. Zhang, H. Yang, J. Wang, J. Long and H. Chang. Mechanical and high-temperature corrosion properties of AlTiCrNiTa high entropy alloy coating prepared by magnetron sputtering for accident-tolerant fuel cladding, *Surf. Coatings Technol.*, 2021, **417**, 127228.

[59] W. J. Shen, M. H. Tsai, K. Y. Tsai, C. C. Juan, C. W. Tsai, J. W. Yeh and Y. S. Chang. Superior oxidation resistance of (Al0.34Cr0.22Nb0.11Si0.11Ti0.22)50N50 high-entropy nitride, *J. Electrochem. Soc.*, 2013, **160**, C531.

[60] X. Li, Z. Zheng, D. Dou and J. Li, Microstructure and properties of coating of FeAlCuCrCoMn high entropy alloy deposited by direct current magnetron sputtering, *Mater. Res.*, 2016, **19**, 802–806.

[61] J. Zhang, J. B. Zhu, Z. Y. Sun and J. C. Li. Preparation of amorphous coatings of AlFeCoNiCuZrV alloy by direct current magnetron sputtering method, *Asian J. Chem.*, 2014, **26**, 5627–5630.

[62] I. Bernardeschi, M. Ilyas and L. Beccai. A review on active 3D microstructures via direct laser lithography, *Adv. Intell. Syst.*, 2021, **3**, 2100051.

[63] R. M. M. Hasan and X. Luo. Promising lithography techniques for next-generation logic devices, *Nanomanufacturing Metrol.*, 2018, **1**, 67–81.

[64] K. Ryu, A. Badmaev, L. Gomez, F. Ishikawa, B. Lei and C. Zhou. Synthesis of aligned single-walled nanotubes using catalysts defined by nanosphere lithography, *J. Am. Chem. Soc.*, 2007, **129**, 10104–10105.

[65] P. Danilevicius, S. Rekstyte, E. Balciunas, A. Kraniauskas, R. Jarasiene, R. Sirmenis, D. Baltriukiene, V. Bukelskiene, R. Gadonas and M. Malinauskas. Micro-structured polymer scaffolds fabricated by direct laser writing for tissue engineering, *J. Biomed. Opt.*, 2012, **17**, 81401–81405.

[66] H. Yin, T. Yamamoto, Y. Wada and S. Yanagida. Large-scale and size-controlled synthesis of silver nanoparticles under microwave irradiation, *Mater. Chem. Phys.*, 2004, **83**, 66–70.

[67] J. Sun, W. Wang and Q. Yue, Review on microwave-matter interaction fundamentals and efficient microwave-associated heating strategies, *Materials (Basel)*, 2016, **9**, 231.

[68] Y. Zhang, P. Chen, S. Liu, L. Fan, N. Zhou, M. Min, Y. Cheng, P. Peng, E. Anderson, Y. Wang, Y. Wan, Y. Liu, B. Li and R. Ruan, Microwave-assisted pyrolysis of biomass for bio-oil production: in *Pyrolysis*, eds. M. Samer, IntechOpen, London, 2017, pp. 129–166.

[69] D. Gupta, D. Jamwal, D. Rana and A. Katoch. in *Woodhead publishing series in biomaterials*, eds. A. M. A. Inamuddin and A. Mohammad, Woodhead Publishing is an imprint of Elsevier Inc, Cambridge, 2018, pp. 619–632.

[70] D. C. Onwudiwe. Microwave-assisted synthesis of PbS nanostructures, *Heliyon*, 2019, **5**, e01413.

[71] M. N. Mustafa, M. A. A. M. Abdah, A. Numan, R. Sulaiman, R. Walvekar and M. Khalid. Microwave-assisted fabrication and optimization of nickel cobalt phosphate for high-performance electrochromic supercapacitors, *J. Energy Storage*, 2023, **73**, 108935.

[72] A. Kubiak and M. Cegłowski. Unraveling the impact of microwave-assisted techniques in the fabrication of yttrium-doped TiO2 photocatalyst, *Sci. Rep.*, 2024, **14**, 262.

[73] C. P. Solution. Wet etching vs. dry etching, https://resources.pcb.cadence.com/blog/2024-wet-etching-vs-dry-etching, (accessed 7 July 2024).

[74] Y. Qiao, D. Wang and J. M. Buriak. Block copolymer templated etching on silicon, *Nano Lett.*, 2007, **7**, 464–469.

[75] H.-D. Yu, M. D. Regulacio, E. Ye and M.-Y. Han. Chemical routes to top-down nanofabrication, *Chem. Soc. Rev.*, 2013, **42**, 6006–6018.

[76] H. Yu, D. Wang and M.-Y. Han. Top-down solid-phase fabrication of nanoporous cadmium oxide architectures, *J. Am. Chem. Soc.*, 2007, **129**, 2333–2337.

[77] T. Fujita, H. Okada, K. Koyama, K. Watanabe, S. Maekawa and M. W. Chen. Unusually small electrical resistance of three-dimensional nanoporous gold in external magnetic fields, *Phys. Rev. Lett.*, 2008, **101**, 166601.

[78] W. Sakuma, S. Yamasaki, S. Fujisawa, T. Kodama, J. Shiomi, K. Kanamori and T. Saito. Mechanically strong, scalable, mesoporous xerogels of nanocellulose featuring light permeability, thermal insulation, and flame self-extinction, *ACS Nano*, 2021, **15**, 1436–1444.

[79] A. Barros, S. Quraishi, M. Martins, P. Gurikov, R. Subrahmanyam, I. Smirnova, A. R. C. Duarte and R. L. Reis. Hybrid alginate-based cryogels for life science applications, *Chemie Ing. Tech.*, 2016, **88**, 1770–1778.

[80] T. D. Gould, A. M. Lubers, B. T. Neltner, J. V. Carrier, A. W. Weimer, J. L. Falconer and J. Will Medlin. Synthesis of supported Ni catalysts by atomic layer deposition, *J. Catal.*, 2013, **303**, 9–15.

[81] S. Wang, Y. Gao, L. Wei, M. Yan, F. Yi, J. Wang, L. Wang, G. liu, A. Song, Y. Li, F. Cai, D. Zhu, D. Xu and Y. Li. Engineering spatial locations of Pt in hierarchically porous KL zeolite by atomic layer deposition with enhanced n-heptane aromatization, *Fuel*, 2023, **337**, 126852.

[82] X. Li, S. Wei, M. Wang, D. Yan, Z. Zhu and D. Tao. Recent progress in regulating of zeolite-based catalysts by atomic layer deposition technology, *J. Fuel Chem. Technol.*, 2024, **52**, 285–292.

[83] M. Saeed, Y. Alshammari, S. A. Majeed and E. Al-Nasrallah. Chemical vapour deposition of graphene – synthesis, characterisation, and applications: A review, *Molecules*, 2020, **25**, 3856.

[84] Y. Liu, T. Wang, J. Wang and W.-P. Pan. Enhancing inward and outward mass transport during chemical vapor deposition of pyrolytic carbon for better synthesis of ZTC, *Microporous Mesoporous Mater.*, 2022, **334**, 111775.

[85] M. Ali, N. Remalli, V. Gedela, B. Padya, P. K. Jain, A. Al-Fatesh, U. A. Rana and V. V. S. S. Srikanth. Ni nanoparticles prepared by simple chemical method for the synthesis of Ni/NiO-multi-layered graphene by chemical vapor deposition, *Solid State Sci.*, 2017, **64**, 34–40.

[86] S. T. Oyama, H. Aono, A. Takagaki, T. Sugawara and R. Kikuchi, Synthesis of silica membranes by chemical vapor deposition using a dimethyldimethoxysilane precursor, *Membranes (Basel)*, 2020, **10**, 50.

[87] H. Perez, V. Jorda, J. Vigneron, M. Frégnaux, A. Etcheberry, A. Quinsac, Y. Leconte and O. Sublemontier, Highly active, high specific surface area Fe/C/N ORR electrocatalyst from liquid precursors by combination of CO_2 laser pyrolysis and single NH_3 thermal post-treatment, *C*, 2019, **5**, 26.

[88] I. I. Lungu, E. Andronescu, F. Dumitrache, L. Gavrila-Florescu, A. M. Banici, I. Morjan, A. Criveanu and G. Prodan, Laser pyrolysis of iron oxide nanoparticles and the influence of laser power, *Molecules*, 2023, **28**, 7284.

[89] R. D'Amato, M. Falconieri, S. Gagliardi, E. Popovici, E. Serra, G. Terranova and E. Borsella. Synthesis of ceramic nanoparticles by laser pyrolysis: From research to applications, *J. Anal. Appl. Pyrolysis*, 2013, **104**, 461–469.

[90] J. Karpiel, P. Lonchambon, F. Dappozze, I. Florea, D. Dragoe, C. Guillard and N. Herlin-Boime, One-Step synthesis of Cu_xO_y/TiO_2 hotocatalysts by laser pyrolysis for selective ethylene production from propionic acid degradation, *Nanomaterials*, 2023, **13**, 792.

[91] X. Zhang, H. Sun, S. Tan, J. Gao, Y. Fu and Z. Liu. Hydrothermal synthesis of Ag nanoparticles on the nanocellulose and their antibacterial study, *Inorg. Chem. Commun.*, 2019, **100**, 44–50.

[92] S. N. Basahel, T. T. Ali, M. Mokhtar and K. Narasimharao. Influence of crystal structure of nanosized ZrO_2 on photocatalytic degradation of methyl orange, *Nanoscale Res. Lett.*, 2015, **10**, 73.

[93] K. Narasimharao and H. S. Kamaluddin. Adsorption of methylene blue and metachromasy over analcime zeolites synthesized by using different Al precursors, *Mater. Today Chem.*, 2023, **32**, 101675.

[94] Q. Zhang, Z. Zhou, Y. Chen, J. Li and H. Kang. Rapid synthesis of SAPO-34 molecular sieve with cheap template for improving MTO performance, *Catal. Commun.*, 2023, **182**, 106751.

[95] J. A. Gómez-Cuaspud, E. Vera-López, J. B. Carda-Castelló and E. Barrachina-Albert. One-step hydrothermal synthesis of LaFeO3 perovskite for methane steam reforming, *React. Kinet. Mech. Catal.*, 2017, **120**, 167–179.

Chapter 3
Different spectroscopy techniques for characterization of materials

3.1 Introduction

Material characterization plays a crucial role in understanding solid materials and exploring the properties of surface reactivity, providing valuable insights into catalytic processes. Various tools can be employed to investigate the distinctive features of materials, thereby uncovering information about their structure, morphology, porosity, and chemical composition. Investigating the surface of materials, especially in terms of its reactivity, means probing into its nature and number of surface sites, any subsequent modification upon functionalization, and the number of adsorbed species as well as potential intermediates in surface-promoted phenomena/reactions.

All technologies share a common principle and typically involve a specific beam impacting the sample, resulting in the emitted beam being detected and analyzed to reveal the information it contains. This process helps obtain the "fingerprint" of solids and/or species or reaction intermediates adsorbed by them. The incident beam can involve various types of radiation such as photons, electrons, ions, neutrons, magnetic, electric, acoustic or thermal fields [1] (see Figure 3.1).

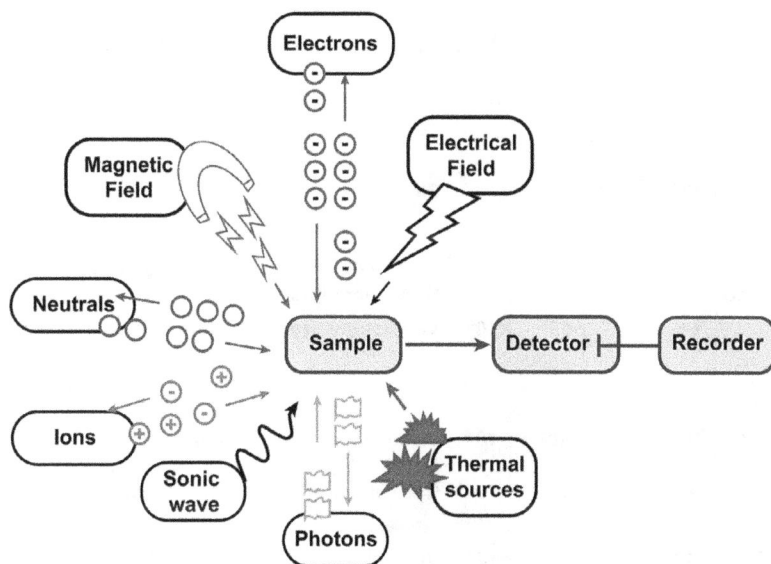

Figure 3.1: Basic phenomena underlying physical techniques. The incident beam is depicted by arrows directed toward the sample, while the emitted beam is indicated by arrows pointing away from the sample (reproduced from Ref. [1]).

https://doi.org/10.1515/9783111316819-003

3.2 Spectroscopy techniques

Spectroscopy investigates the interaction between light and materials, such as molecules or atoms, through the absorption of electromagnetic radiation. The word "spectroscopy" refers to a broad range of techniques which utilize electromagnetic radiation to investigate about the structure and properties of matter. Spectroscopy techniques use different types of electromagnetic radiation to interact with matter, probing certain aspects of a sample to learn about its consistency or structure. Different type of electromagnetic radiation, with different energy can be used to examine various molecular properties. The study of how electromagnetic radiation interacts with matter as a function of the radiation's wavelength or frequency is known as spectroscopy. The spectroscopy experiments involve exposing an electromagnetic radiation of a specific energy from the source through the samples containing substances of interest, causing in absorption or emission of energy. The underlying principle behind the spectroscopic techniques is to expose the sample to a beam of electromagnetic radiation and study how sample reacts to it. Spectroscopy methods are commonly used techniques to identify and explain the elements and compounds of atoms and molecules in the samples. They are determined by evaluating the radiant energy absorbed or emitted by the sample or object. A beam of electromagnetic radiation such as X-rays, infrared (IR) rays, UV rays, and so on is passed on the sample, and the response of the sample is measured using the wavelength of the electromagnetic spectrum applied from an external energy source.

Types	E (kJ mol^{-1})	ν (s^{-1})	λ (m)	Incident beam	Technique
Radio	10^{-5}	10^7	10^1	Nuclear spin flip	NMR
Microwave	10^{-3}	10^9	10^{-1}	Electron spin flip	EPR
Infrared	10^{-1}	10^{-1}	10^{-3}	Molecular rotaion	IR, Raman
Near-Infrared			10^{-5}	Molecular vibration	
Visible	10^3	10^3	10^{-7}	Valence electron transition	UV-vis
Ultravoilet					
Vacuum UV	10^5	10^5	10^{-9}	Valence electron excitation	UPS
X-Ray	10^7	10^7	10^{-11}	Core electron excitation	XPS, XAS
Gamma-Ray	10^9	10^9	10^{-13}	Nuclear transition	Mossbauer

Figure 3.2: Principal spectroscopic techniques and the corresponding phenomena associated with the incident beam, based on its specific characteristics.

By measuring the absorption or emission of electromagnetic radiation by the sample and these techniques provide information about the molecular structure, energy levels, and chemical composition of the material. The obtained information can be used to identify and quantify different types of chemical species and to investigate their behavior and interactions. Spectroscopy techniques include Absorption Spectroscopy, Emission Spectroscopy, Fluorescence Spectroscopy, Infrared Spectroscopy, Raman Spectroscopy, and Ultraviolet-Visible Spectroscopy, among others. Electromagnetic radiation waves combine electric and magnetic fields with varying energy sources, as illustrated in Figure 3.2.

3.3 Fourier transform infrared (FT-IR)

Infrared (IR) spectroscopy is a non-destructive technique employed to gather information about the morphological and physico-chemical characteristics of catalytic systems. Since the 1960s, IR spectroscopy has been utilized to study powdered heterogeneous catalysts, and in situ studies of working catalysts have been conducted since the 1970s. IR spectroscopy experienced a renewal with the advent of Fourier transform (FT) instruments in the 1980s, significantly increasing the technique's sensitivity and spectral quality. By the 1980s, FT-IR spectrometers became commonplace in chemistry laboratories. Since then, their cost and the required hardware for mathematical analysis have consistently decreased. Currently, most heterogeneous catalysis laboratories are equipped with an IR spectrometer, a fundamental technique for characterizing heterogeneous catalysts. FT-IR methods determine the amount of light that a sample absorbs at each wavelength by shining a monochromatic light beam on it [2–4].

3.3.1 FT-IR concept

The IR region is divided into subregions as shown in Table 3.1. Vibrational spectroscopy studies in the mid-IR region encompass the electromagnetic spectrum in the range from 25 to 2.5 μm, or, in wavenumber terms, from 400 to 4,000 cm^{-1}. The frequency of the radiation is designated by v, and its wavelength is expressed as follows:

$$\lambda = c/v \tag{3.1}$$

$$v = c/\lambda \tag{3.2}$$

where c represents the speed of light (3×10^{10} cm/s) and λ is a wavelength. The unit employed in IR spectroscopy is the wavenumber. It is defined as the number of cycles observed per centimeter, hence the unit wavenumber is expressed as cm^{-1}, also known as frequency. The quanta associated with radiation's energy (E) value is directly proportional to its wavenumber (v):

$$(E) = h \cdot v = h \cdot c / \lambda \tag{3.3}$$

where h is the Planck constant ($6.6 \times 10-34$ Js).

Table 3.1: Range of IR radiation.

	Near-IR	Mid-IR	Far-IR
Wavelength (λ/μm)	1–2.5	2.5	50–1,000
Wavenumber (λ^{-1}/cm^{-1})	10,000–4,000	4,000–200	200–10
Energy (eV)	1,240–0.496	0.496–0.025	0.025–0.0012

Chemical bonds in molecules undergo various types of fluctuations – changes in bond lengths or angles. These vibrations contribute to the overall energy of the molecules, and since molecule energy can be quantified, molecules have a certain number of energy sub-levels. These energy sub-levels can be calculated using equations based on the principles of quantum mechanics:

$$E = (n + 1/2) \cdot h \cdot v_1 \tag{3.4}$$

Here, n denotes a quantum number (0, 1, 2, 3, etc.) corresponding to different energy levels. According to a selection principle, transitions between distinct energy levels may be allowed or forbidden. Only transitions to the next energy level are allowed, resulting in molecules absorbing energy equal to hv (where h is Planck's constant, and v is the frequency). This principle is not strictly followed; sometimes, transitions can occur at $2\,hv$, $3\,hv$, or higher frequencies (Figure 3.3) [5].

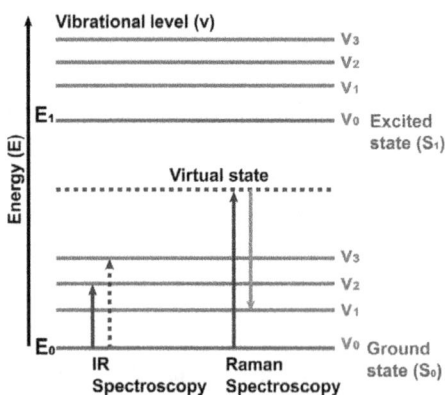

Figure 3.3: Jablonski diagram showing the energy transitions in IR and Raman spectroscopy techniques.

The electromagnetic radiation in the IR region has relatively weak energy (longer wavelengths), causing molecules to only vibrate when absorbed in this range. For molecules with more than two atoms, the molecular vibration can be classified into two categories: (i) stretching (v) and (ii) bending (δ). Stretching vibration leads to changes in bond lengths, while bending vibration results in alterations in angles between bonds. For smaller molecules or specific groups of atoms within larger molecules, stretching vibrations can be further classified into (i) symmetric (characterized as simultaneous lengthening or shortening of bonds) and (ii) antisymmetric (involving some bonds elongating while others in the same group of atoms shorten). Bending vibrations, especially in cases where one portion of a molecule is larger than another, can be divided into (i) in-plane bending and (ii) out-of-plane bending; in-plane bending subsequently divides into scissoring and rocking, while out-of-plane bending incorporates wagging and twisting. Furthermore, vibrations involving simultaneous changes in the lengths or valence angles of multiple bonds are referred to as skeletal frequencies. This phenomenon can be visualized in Figure 3.4. Vibrational spectra of a compound represent the outcome of either direct absorption of IR radiation, resulting in a change in the vibrational state of molecules (IR spectroscopy) [5].

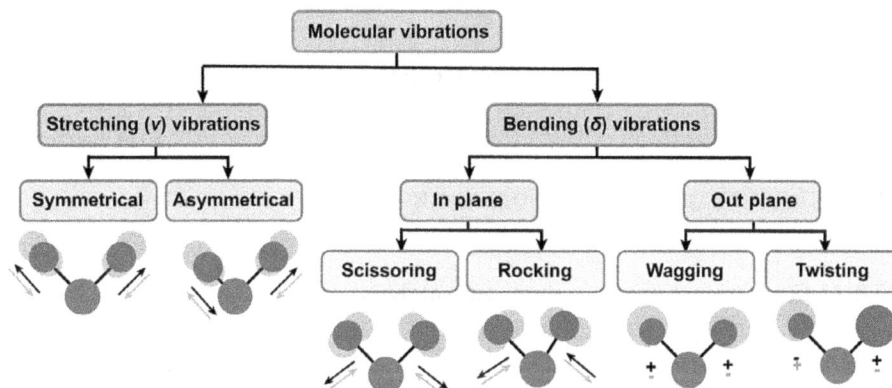

Figure 3.4: Phenomenological nomenclature of molecular vibrations.

3.3.2 FT-IR instrument

IR spectrometers consist of a radiation source, a system for discriminating between different wavelengths, a sample stage, and a detector. Within these instruments, wavelength differentiation is typically achieved using FT technology in research associated with catalysis because, in the IR range, grating monochromators are primarily utilized for precise, high-end measurements. The Michelson interferometer, the cornerstone of FT spectrometers, involves splitting radiation using a semitransparent mirror (beam splitter, Figure 3.5). Subsequently, the beams are reunited after traveling respective paths, including one mirror that moves in relation to the initial partial beam in order to manipulate its optical path, denoted as δ.

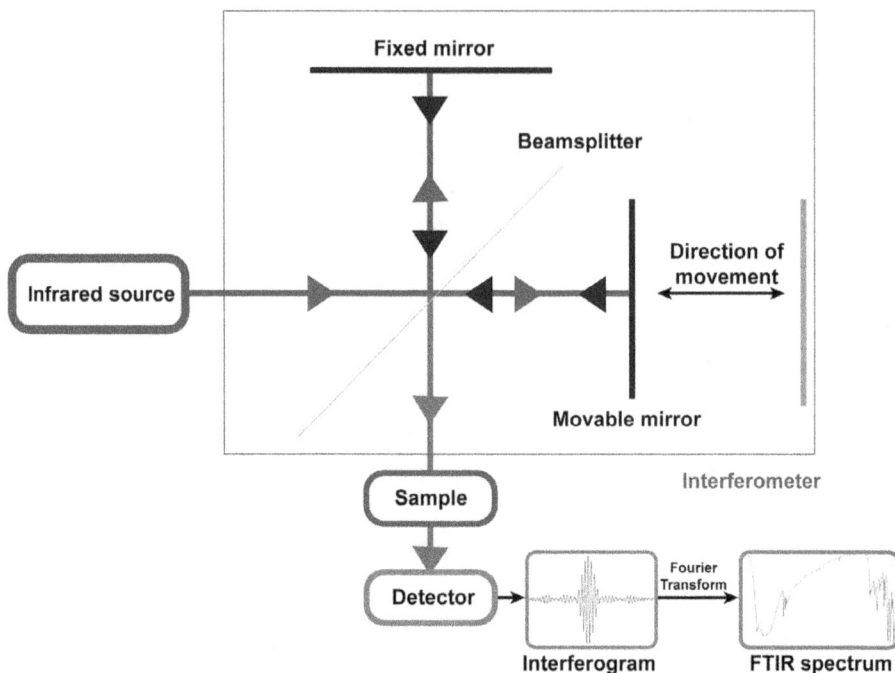

Figure 3.5: Interferometer schematic in FT-IR spectrometer.

3.3.3 FT-IR spectroscopy in heterogenous catalysis

IR spectroscopy offers insights into six aspects: (i) surface functional groups, (ii) the nature and composition of surface and/or adsorbed species, (iii) the intensity of chemical bonds, (iv) reaction mechanisms [6], (v) lattice vibrations which typically occur below 1,200 cm^{-1} for most solids to ascertain structures, and (vi) adsorbate materials are used to investigate the surface characteristics of catalysts such as acidity, basicity, and oxidation states. An IR spectrum of molecule is based on absorption of IR radiation with wavelengths within the mid-IR range. When molecules vibrate, they change IR absorption, forming specific absorption bands known as molecular fingerprints. These bands enable the detection of compounds and provide structural details, making IR spectroscopy a powerful tool [7].

In general, potassium bromide (KBr) is commonly used to disperse catalyst materials (typically at a percentage of 0.5–1%). By compressing the resulting powder (at a pressure of 5–8 tons/cm or 5–8 × 10^8 Pa) into a pellet and placing it directly in the IR beam, this method can generate valuable insights into the structural characteristics of materials.

3.3.3.1 Zeolite catalysis

The main bands observed for zeolites typically fall in the 400 and 1,300 cm^{-1} regions. The vibrational spectrum of zeolites derives from several components: **(i)** vibrations from the zeolite framework (denoted as v(T-O)), **(ii)** those originating from charge-balancing cations, and **(iii)** O-H groups present on the surface of the zeolite.

(i) Framework vibration

The zeolite framework is constructed from a network of TO$_4$ units (T = Si or Al). The vibrational modes of the zeolite framework give rise to distinct bands in the mid-IR region. According to Flanigan and Khatami [8], in the IR spectra of zeolites within the 1,250–300 cm^{-1} range, two major vibrational modes are commonly detected: external and internal.

The internal vibrations within TO$_4$ tetrahedron tend to be relatively insensitive to the specific framework structure of zeolitic materials. Conversely, the external vibration modes, particularly those in the 1,050–1,150 cm^{-1} region, are highly sensitive to the topology (arrangement of TO$_4$ atoms) and can exhibit distinct patterns based on the types of building units present (e.g., double six-ring and 12-ring). The removal of aluminum from tetrahedral sites should result in noticeable changes in both vibration classes. The most pronounced absorption typically occurs within the 950–1,050 cm^{-1} region and is associated with the asymmetric T-O stretching vibration (where T denotes either Si or Al). This vibration is highly sensitive to the Si/Al ratio, with a shift toward higher frequencies as Si content increases [3]. Table 3.2 provides an example of the predominant bands observed in zeolite Y [4], with a visualization in Figure 3.6.

(ii) Cation vibrations

The positions of the corresponding IR bands associated with cations are influenced by their charge, mass, and the specific interaction they have with the zeolite structure. The NH$_4^+$ form, for instance, exhibits a distinct band at approximately 1,460 cm^{-1} attributable to the departing ammonium group.

(iii) Hydroxyl group vibrations

Hydroxyl groups originating from water molecules typically contribute to the v(O-H) stretching and δ(OH) bending vibration bands, which appear as broad features in the spectrum around 3,600 cm^{-1} (O-H stretch) and approximately 1,630 cm^{-1} (OH bending). During the activation of certain sensitive zeolites, such as mordenite and aluminum-rich varieties, caution is advised as they may undergo damage. This process can lead to the formation of an extra-framework phase within the solid. This assessment can be performed after activation by determining the following: the absence of AlOH [v(OH)] vibrational bands around 3,660 cm^{-1} and, conversely, the determination of the specific Lewis sites (measured via pyridine adsorption).

Table 3.2: Zeolite band assignments [3].

Internal tetrahedra	950–1,250 cm^{-1}	Asymmetrical stretch (n_{asym}) O-T-O
	650–720 cm^{-1}	Symmetrical stretch (n_{asym}) O-T-O
	420–500 cm^{-1}	T-O bend
External linkages	1,050–1150 cm^{-1}	Asymmetrical stretch (n_{asym}) O-T-O
	750–820 cm^{-1}	Symmetrical stretch (n_{sym}) O-T-O
	500–650 cm^{-1}	Double ring (D n-member ring)
	300–420 cm^{-1}	Pore opening vibration
OH groups	3,745 cm^{-1}	Vibration of isolated silanol groups, like those observed in silica, very weakly acidic
	3,610 cm^{-1}	Bridged hydroxyl group (SiOHAl), Brønsted acid sites
	3,630 cm^{-1}	High-frequency (HF) band for OH in the supercages
	3,550 cm^{-1}	Low-frequency (LF) band for OH groups in the sodalite unites or hexagonal prisms of the structure
Present of extra-framework, that perturb the OH groups	3,600 cm^{-1}	Perturbed HF band bearing the strongest acidity (so-called enhanced acidity)
	3,520 cm^{-1}	Perturbed LF band, probably also enhanced acidity, but out of reach for most molecules, and therefore without any direct role in acidity or catalysis
Nonacidic OH groups	3,660–3,680 cm^{-1}	Aluminic OH (extra-framework phase)
	3,775–3,785 cm^{-1}	Aluminic OH groups, often described as basic

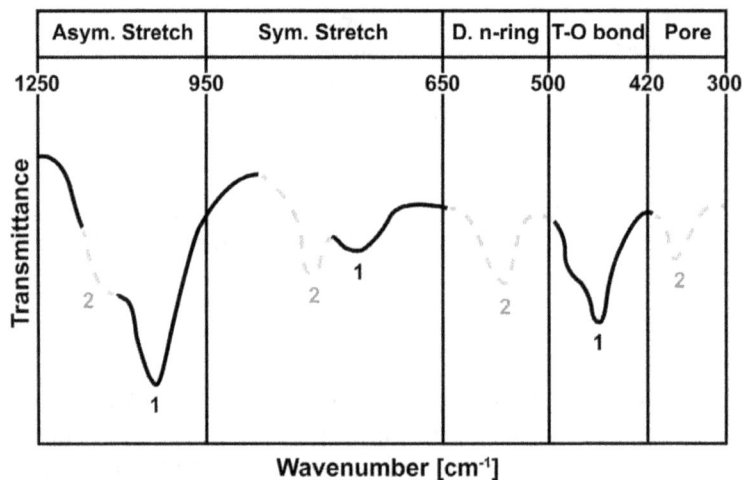

Figure 3.6: Infrared assignments illustrated with Y zeolite spectrum. (1) Full lines: internal TO$_4$: structure insensitive. (2) Dashed lines: external TO$_4$ linkages: structure sensitive (reproduced from ref [4]).

For instance, when analyzing the vibrational spectrum of zeolite Y material (Figure 3.7a), specific bands can be identified: at 3,410 cm^{-1}, corresponding to vibrations associated with Si-OH, Si-OH-Al, and OH hydroxyl groups; at 574 cm^{-1}, attributed to the external linkage of double rings; at 982 cm^{-1}, representing Si-O, SiO-Al, Al-OH asymmetric and symmetric stretching vibrations linked to the internal T-O_4 framework (where T can be either Si or Al); and at 1,146 and 790 cm^{-1} are attributed to asymmetric and symmetric stretching vibrations linked to the external T-O_4 structure. These are the distinguishing features of zeolite Y material. The band observed at 1,389 cm^{-1} is assigned to the external linkage related to the double-ring structure inherent in the faujasite (FAU) framework [9]. The FT-IR analysis could be used to quantify the amount of the adsorb species on the surface of materials by investigating the changes in the intensity of certain of IR bands which are directly proportional to the amount of adsorbed species accumulate on the surface. Figure 3.7b illustrates discrete peaks associated with functional groups prior and after the adsorption of NH_4^+ ions. Before adsorption, zeolites exhibit vibrations within the 600–798 cm^{-1} range, attributed to the deformation band of T-O, resulting from internal vibrations of (Si, Al)O_4 tetrahedral units in the zeolite framework. Additionally, the frequency spectrum in the range of 1,012–1,052 cm^{-1} is ascribed to T-O-T stretching bands occurring at the zeolites surface. The peak observed at 1,628 cm^{-1} experiences a shift to 1,636 cm^{-1}, representing bending vibrations of water molecules adsorbed onto the zeolite surface. Following adsorption of NH_4^+ ions, two new peaks emerge at 1,400 and 3,447 cm^{-1}. The former is related to C-H bending vibrations, while the latter corresponds to O-H stretching modes of the hydroxyl group [10].

3.3.3.2 Metal and metal oxide

IR analysis was employed to characterize surface functional groups, emphasizing hydroxide (OH) groups present on oxide catalysts and support materials, which exhibit different structural environments. Specific data are provided in Table 3.3.

FT-IR spectra were utilized to investigate the impact of noble metals (e.g., Au and Ag) deposition on LaMnO$_3$ nanocomposites. The corresponding FT-IR spectra for both bulk and noble metal-deposited LaMnO$_3$ samples are presented in Figure 3.8. Specific absorption bands were identified at 643, 854, 985, 1,078, 1,473, and 1,624 cm^{-1} for all specimens. It is indicated that distinctive absorption features associated with LaMnO$_3$ nanoparticles appear at around 600, 820, 1,100, 1,380, 1,450, and 1,650 cm^{-1} for LaMnO$_3$ nanoparticles. The primary IR absorption band within the range of 615–643 cm^{-1} originates from the stretching mode of the Mn–O–Mn bonds linked to the octahedral MnO$_6$ unit. This vibrational feature, representative of the ABO$_3$ perovskite structure, attests to the successful synthesis of LaMnO$_3$. The bands occurring near 1,624 cm^{-1} are believed to correlate to bending vibrations of N–H bonds (present in secondary amines), while the bands around 1,380 and 1,450 cm^{-1} are attributed to bending vibrations in N–O bonds (associated with nitrate groups). As indicated in the FT-IR spectrum in Figure 3.8, the deposition of noble metals results in a subtle shift in the band positions [11].

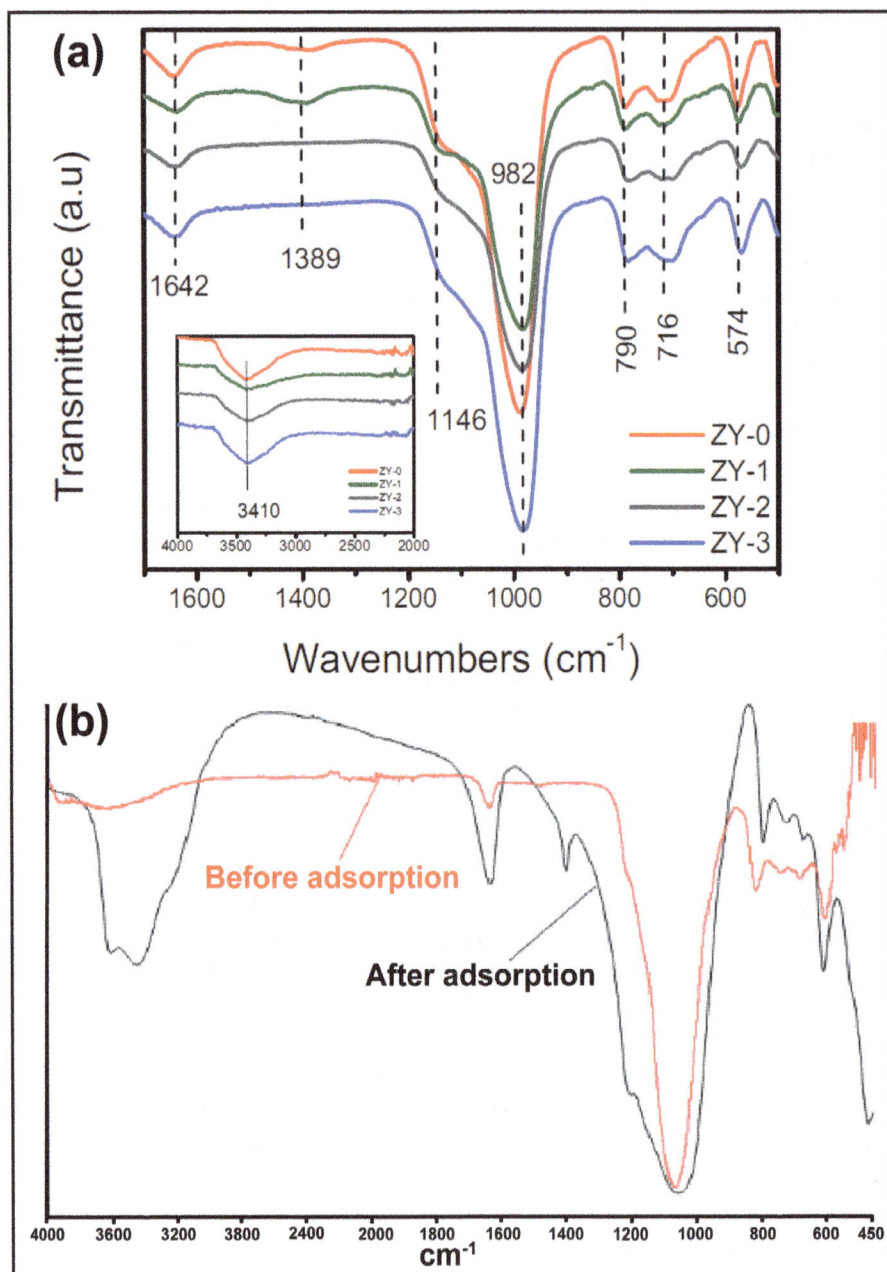

Figure 3.7: Infrared assignments illustrated for **(a)** Zeolite Y and it is modified counterparts (adapted from ref. [9]) and **(b)** synthesized zeolite before and after the adsorption of NH_4^+ ions (adapted from ref. [10]).

Table 3.3: Band assignments for different species.

Residual impurities	Below ~1.600 cm^{-1}	They characterize the presence of remaining impurities (such as carbonate, nitrate, and sulfate species) when metal oxide after activation process at 500 °C
Carbonates	~1,415 cm^{-1}	The contact of the CO_2 present in air with metal oxide basic sites, such as oxygen or OH anions, leads to carbonate or hydrogencarbonate groups, respectively
	100 cm^{-1}	Adsorbed Uni- or monodentate carbonates
	300 cm^{-1}	Adsorbed bidentate carbonates
Nitrates	1,380 cm^{-1}	Adsorbate nitrates for free anion
	~1,050 cm^{-1}	Adsorbate nitrates
Sulfates	~1,340 cm^{-1}	Characterized for surface sulfate species of specific (S=O) group on activated oxides through three S-O-M linkages together with several bands below 1,260 cm^{-1}
OH groups	3,500–3,800 cm^{-1}	Elongation vibration in OH group v(OH)
	1,200 and 2,000 cm^{-1}	The δ(OH) vibration is a deformation of the OH bond parallel to the plane of the two bands on the oxygen atom (in-plane deformation).
	Below 1,500 cm^{-1}	The Y(OH) vibration is a deformation of the OH group out of the plane of the bonds on the O atom
	3,745 cm^{-1}	The v(OH) frequency for a basic oxide such as MgO, for a weakly acidic amorphous silica, and for a strongly Brønsted acid such as amorphous silica–alumina
	3,790, 3,770, 3,755, 3,730, and 3,680 cm^{-1}	The v(OH) group adsorbed on alumina

3.3.4 FT-IR using probe molecules

FT-IR spectroscopy can be utilized to characterize the acidity and basicity of solid catalyst materials through the use of probe molecules. Commonly used bases molecules such as amines, pyridines, ammonia, carbon monoxide, to investigate the acidity of catalysts. The characterization of basic sites in catalysts typically involves using acidic probe molecules, which often leads to more complex interactions. Acidic probe molecules include carbon dioxide (CO_2), protonic molecules such as acetylene, propyne (methylacetylene), butyne (ethylacetylene), pyrrole (C_4H_4NH), and chloroform ($CHCl_3$) as well as methanol (CH_3OH).

Pyridine: This compound serves as one of the most commonly utilized probes in the study of solid acids. Its use is a straightforward process, providing insights into the types and concentrations of sites present on the surface. Pyridine is recognized as a very strong base, capable of being protonated on Brønsted acid sites (BASs), and also

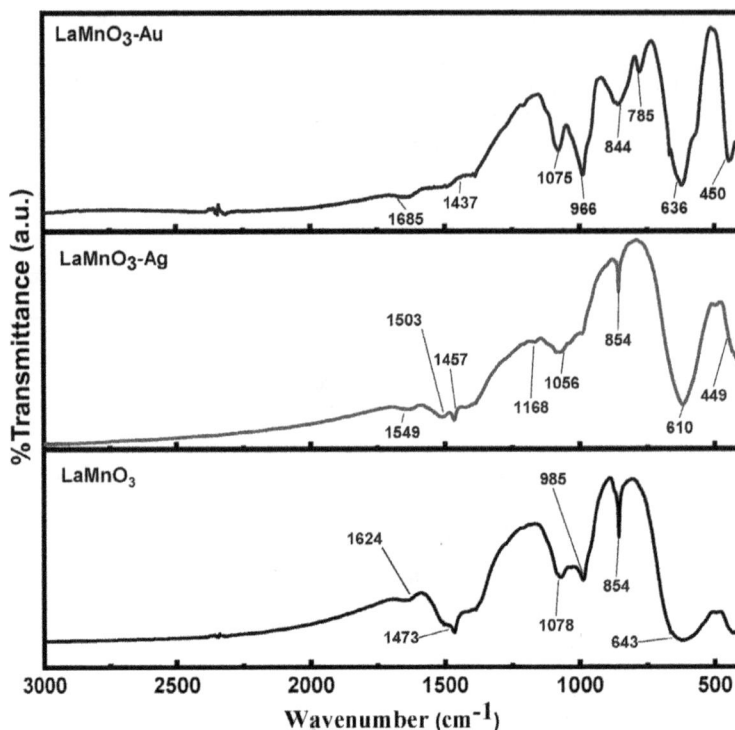

Figure 3.8: FT-IR spectra for the bulk LaMnO$_3$ and noble metals (e.g., Au and Ag) deposited of nanocomposites (adapted from ref. [11]).

displays strong coordination on Lewis acid sites. Weak BASs such as the silanol group present in silica are capable of forming hydrogen bonds with pyridine. Three distinct complexes can form on the surface of an oxide material: PyH$^+$ (protonated pyridine), PyL (pyridine coordinating with the acidic Lewis site), and Py–HO (hydrogen-bound pyridine, along with physiosorbed pseudo-liquid pyridine). These complexes can be identified through the vibrational frequency of the heterocyclic ring, which falls within the range of 1,400–1,700 cm^{-1}.

Acetonitrile: It is a stronger base than carbon monoxide (CO), and at room temperature, it can adsorb on both Lewis and BASs in oxides. Its small molecular size allows it to access most of the acidic sites (Table 3.4). The strong basicity of acetonitrile makes it an interesting probe as it forms the strongest possible hydrogen bonds with Brønsted sites before the actual proton transfer takes place. When acetonitrile interacts with Brønsted sites, the strongest hydrogen bond is formed, and the largest Dn(OH) changes are observed.

Carbon dioxide: is widely used as a probe molecule for characterizing basic sites. When it reacts with basic hydroxy groups, it forms surface hydrogen carbonates, or when it reacts with oxygen anions, it forms carbonates. The structures of carbonate species formed depend on the specific environments of basic sites. The formation of hydrogen carbonates indicates the basic properties of high-frequency OH groups on beta zeolites that are attached to Lewis acidic Al atoms, with bands around 3780 cm^{-1}. CO_2 is not a reliable probe for directly assessing basic strength or for accurate quantification of basic sites [12].

Pyrrole: is a useful probe for studying basicity because it contains an NH group, and can form hydrogen bonds with basic oxygen atoms on oxides and zeolites, leading to a shift in the v(NH) vibration band. Pyrrole can adsorb either without dissociating (PYH) or dissociating , when interacting with more basic adsorption sites on the surface of the material), leading to the formation of pyrrolate ions (PY$^-$). The interaction of pyrrole (PYH) with surface hydroxy groups on the material's surface leads to the formation of NH—(OH) hydrogen-bridged species between the pyrrole nitrogen and the hydroxy oxygen atom. These hydrogen bonds can be either monodentate or bidentate, depending on whether the hydroxy groups are attached to the surface with one or two covalent bonds, respectively. When (PY$^-$)bound to surface hydroxy groups through a polarizable hydrogen bond, the pyrrolate anion may take on a non-planar structure, resulting in complex progressions of IR bands involving CH-stretching modes and ring deformation vibrations. When pyrrole adsorbs without dissociating, through an NH...O bridge with O^{2-} moderately basic surface centres, the shift of the NH-stretching frequency reflected the O^{2-} basicity (Table 3.4) [13].

Table 3.4: Typical frequencies (cm^{-1}) for bands after probes adsorption [12, 13].

	Silanol groups	Brønsted acid sites	Lewis acid sites
After CO adsorption			
v(CO)	~ 2150	2155–2180	2190–2240
Δv(CO)	~ 15	20–45	50–100
Δv(OH)	~ 90	100–400	
After acetonitrile adsorption			
v(C≡N)	~ 2285	2295–2300	2325–2330
Δv(C≡N)	~ 40	50–55	~ 85
Δv(OH)	~ 300	500–1000	

After PYH/PY$^-$ adsorption			
	Symmetry	C$_4$H$_4$NH (PYH)	C$_4$H$_4$N$^-$ (PY$^-$) solid
		Gas / Solid	
δ(NH)	Antisymmetric (B)	960[a] ---	---
v(NH)	Symmetric (A)	3527 ---	---
δ(CH)	Antisymmetric (B)	1049[a] ---	---
v(CH)	Symmetric (A)	3148, 3125 3129, 3113	3074, 305

Table 3.4 (continued)

	Symmetry	C₄H₄NH (PYH)		C₄H₄N⁻ (PY⁻) solid
		Gas	Solid	
	Antisymmetric (B)	3140, 3116	3123, 3105	3090, 305
ν(ring)	Antisymmetric (B)	1521, 1424	1522, 1432	1453, 1361
	Symmetric (A)	1472, 1391	1472, 1379	1442, 1344

[a]δ(CH), δ(NH) vibrational coupling.

3.4 Raman spectroscopy

Raman spectroscopy, a complementary technique discovered in 1928, gained considerable prominence in the 1960s with the advent of lasers, and further advancements in the 1980s and 1990s came with the emergence of charge-coupled device detector (CCD) arrays and high laser sources. Additionally, the development of narrow-band laser line rejection filters enabled the incorporation of filter sets and single optical gratings, effectively addressing the significant signal loss associated with conventional triple monochrome systems. Furthermore, the implementation of multichannel signal detection resulted in markedly reduced acquisition times, significantly improving the signal-to-noise ratio. The combination of these technological advancements paved the way for the introduction of new Raman spectrum microscopes in the 1990s, making them a relatively affordable benchtop laboratory tool capable of complementing conventional IR spectrum microscopes.

IR and Raman spectroscopy share a common principle, involving incident radiation interacting with molecular vibrations to generate distinctive spectra (Figure 3.3). However, there is a fundamental difference in the interaction mechanisms. In IR spectroscopy, the radiation is absorbed, inducing vibrational state transitions, whereas in Raman spectroscopy, a scattering technique is employed, resulting in the incident radiation coupling with the molecule's vibration polarization and producing or annihilating vibrations. This produces a distinct Raman spectrum (Figure 3.9).

The underlying principles of IR and Raman spectroscopy are different, providing a complementary approach to material analysis. For a vibrational mode to be active in IR spectroscopy, a dipole change is necessary. In Raman spectroscopy, it is the polarization change that is directly observed. Thus, asymmetric polar bonds typically exhibit strong vibrational modes in IR spectra, while Raman spectroscopy excels in probing non-symmetric polar groups. For example, the O-H vibrational mode of water is strongly active in IR spectra, but weakly apparent in Raman spectra, making Raman spectroscopy particularly valuable for in vivo biomedical applications. The alternative origin of these techniques leads to another difference, where IR directly measures the absorption of IR radiation, while Raman scattering is also applicable in the ultraviolet, visible, or near-

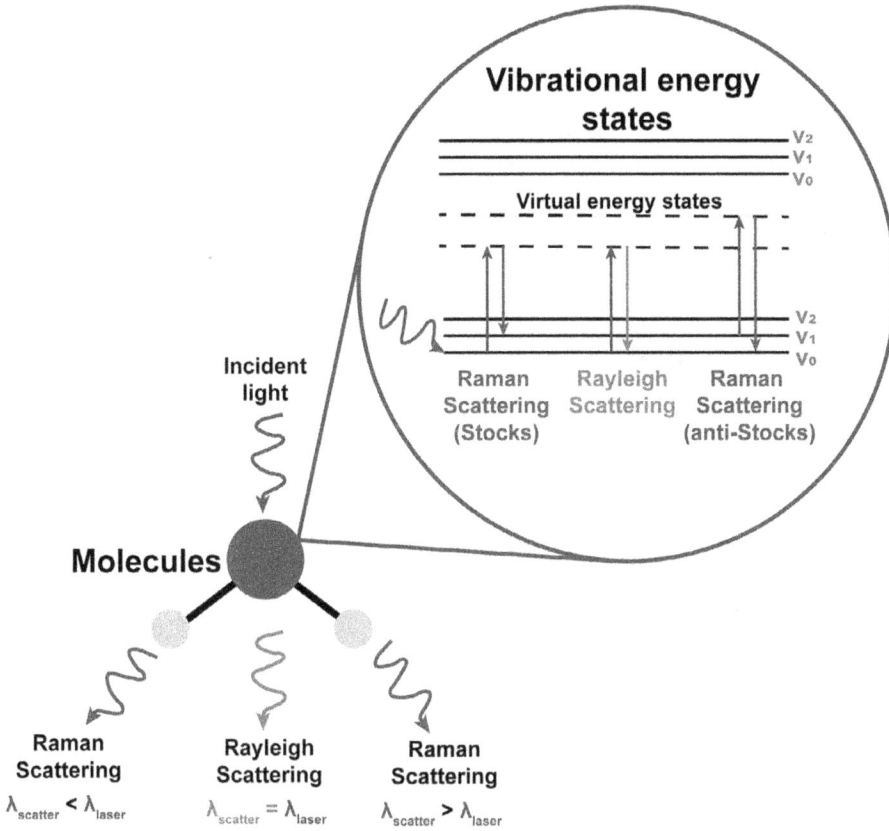

Figure 3.9: Depiction of light scattering by vibrating polarization.

infrared (NIR) regions of the spectrum. This feature allows for higher spatial resolution mapping or profiling with inherent Raman scattering, as well as a stricter wavelength determination within the diffraction limit (~1 μm for Raman, 5–10 μm for IR). However, many applications prefer using NIR as the Raman analysis source to minimize interference from sampling dispersion, fluorescence, or photodegradation.

3.4.1 Raman instrument

A standard Raman spectrometer typically consists of three main components: the excitation source, the sampling device, and the detector. Over the years, these three components have evolved in various forms, but modern Raman instruments are now primarily based on lasers as the source of excitation, with detector spectrometers, microscopes, or fiber optic probes utilized for sampling devices.

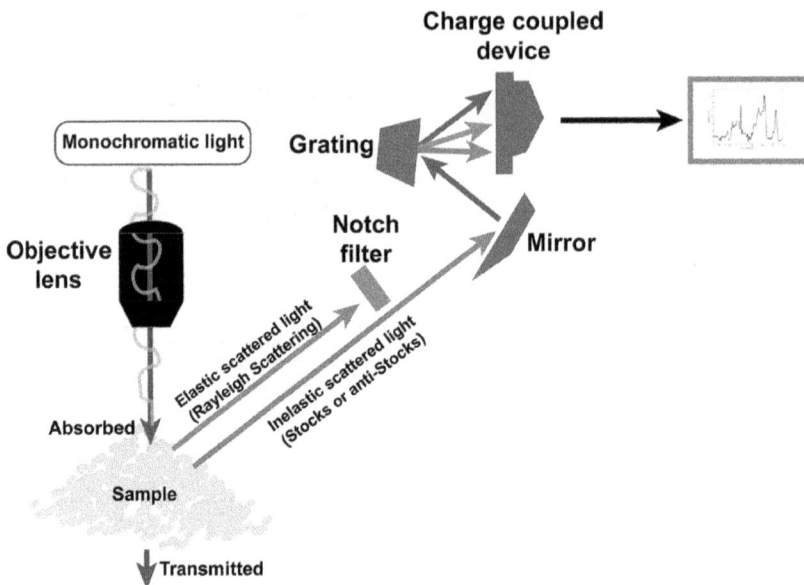

Figure 3.10: Typical instrumentation for Raman microscopy.

Figure 3.10 illustrates a schematic diagram for a typical Raman system. Laser sources provide consistent and powerful beams for excitation. Various lasers can serve as the light source. Bandpass "interference" filters are used to clean up the laser spectrum and remove plasma lines. Dispersive instruments utilize notch filters and high-quality grating monochromators. Double or triple coating monochromators, rejection filters, super notch filters, holographic notch or edge filters, or holographic filters are employed to isolate weak Raman lines from the intense Rayleigh-scattered radiance. Charge transfer devices (CTDs), such as CCDs and charge injection devices, serve as detectors in array form. CTD arrays function by converting incoming optical signals into charge, which is then integrated and transmitted to readout devices. CTDs are typically manufactured using silicon to detect laser wavelengths below 1 μm, whereas laser wavelengths above 1 μm utilize single-element detectors based on low-bandgap semiconductors, such as germanium (Ge) or indium-gallium-arsenic (InGaAs). A diffraction grating is employed to disperse the light, and an array of discrete detectors placed behind the grating produces an optical signal that can be measured and converted into Raman scattering with a known distribution of intensities as a function of wavelength, determining the spectrum resolution. Additional factors that significantly impact spectral resolution include the wavelength, with shorter wavelengths typically providing enhanced spectral resolution. Additionally, the length of the spectrometer, defined by the distance between the grating and the detector, plays a role, as a longer distance generally leads to higher spectral resolution. The objective lens serves dual functions, acting as both the source and collector of light. It directs incident light to the sample, while it also collects and transmits dispersed light. In backscattering geometry, Raman scattering is captured by the objective lens and directed to the grating for dispersion and detection. The dispersed and detected spectral light is displayed as a Raman shift from the source wavelength, converted to a wave number unit (cm^{-1}), facilitating straightforward comparisons and contrasts between the Raman spectrum and the equivalent FT-IR spectrum.

3.4.2 Raman spectroscopy in catalysis

Raman spectroscopy reveals characteristic vibration modes specific to zeolite structures. Most zeolites exhibit pronounced peaks between 200 and 1,200 cm^{-1} in the spectral range (see Figure 3.11). Table 3.5 summarizes the Raman bands for different zeolite group structures, including Natrolite (fibrous), Chabazite (six-cyclic ring), Analcime (single-connected four-ring chains), Gismondine (doubly connected four-ring), Mordenite, and Heulandite [14].

Table 3.5: Typical frequencies (cm^{-1}) for bands after probes adsorption.

	Wavenumbers
T–O–T (T = Si and Al, O = oxygen) modes	In the range 379–538 cm^{-1}
The M–O (M = metal) modes	In the range 250–360 cm^{-1}
T–O bending modes	In the range 530–575 cm^{-1}
The analcime group (with four-ring chains) T–O–T modes	In the 379–392 cm^{-1} and 475–497 cm^{-1}
The gismondine group (with four-ring chains) T–O–T modes	In the 391–432 cm^{-1} and 463–497 cm^{-1}
The chabazite group (with six-cyclic ring) M–O modes	In the 320–340 cm^{-1}
T–O–T modes	In the 477–509 cm^{-1}
The mordenite group	In the 397–410 cm^{-1} and 470–529 cm^{-1}
The heulandite group	402–416 cm^{-1} and 480–500 cm^{-1}

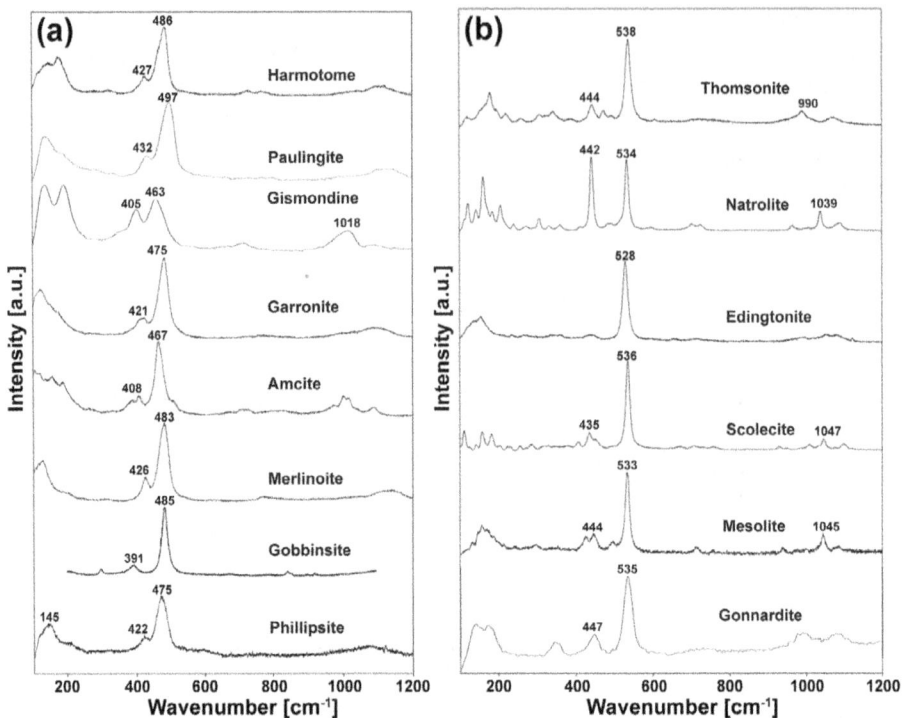

Figure 3.11: Raman spectra of (a) gismondine group and (b) natrolite group zeolites (adapted from ref. [14]).

3.5 Electron transitions

In UV and visible regions, the electron transitions can occur within molecular or complex orbitals (Figure 3.12a). Transitions can also occur between bands and atomic states (for cations) in solid materials (Figure 3.12b). Additionally, photons in this range can excite surface plasmons in metals, resulting from collective vibrations of free electrons relative to atomic nuclei.

Molecular species transitions are commonly studied in liquid phases to examine the processes of precursor solutions. Additionally, these transitions can be used to investigate carbonaceous and adsorption deposits on solid surfaces. Molecular electrons can be excited from occupied bonding (σ, π) and non-bonding (n) orbitals to unoccupied states (e.g., π^* or σ^*). These transitions, known as HOMO-LUMO (highest occupied molecular orbital (HOMO) to lowest unoccupied molecular orbital (LUMO)) transitions, can be further distinguished by spin orientation (single or triple) or symmetry, among other attributes. π-π^* transitions in molecules and structures with conjugated double bonds are common examples of this phenomenon. The wavelength (λ) and absorption coefficient (α) of electronic transitions in molecules tend to increase as the size of π systems

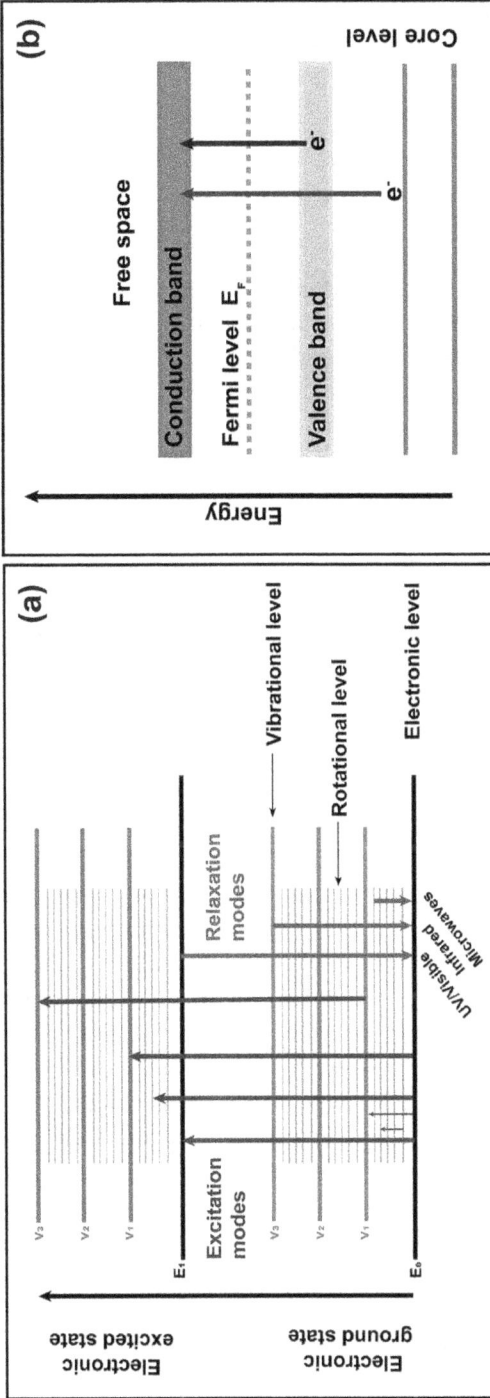

Figure 3.12: The electron transition **(a)** between the orbits of molecules in solution and **(b)** between bands in solid material.

expands. Furthermore, UV-vis signals are expected from molecules that contain specific functional groups such as –C=O (carbonyls), –N=O (nitriles), or atoms with incomplete valence shells.

3.5.1 Measuring UV-vis spectra

Ultraviolet and visible light could interact with matter through scattering, absorption, and specular reflection. The light absorption depends on wavelength of the light and nature of the sample and can be measured by determining the amount of light that is absorbed or transmitted or reflected by the sample. The absorbed light energy could be converted into thermal energy or fluorescent radiation in some cases. In the process of determination of amount of light absorbed or transmitted by the sample, it is usually assumed that scattering and specular reflection are low or negligible. This assumption applies primarily to liquid samples, where reflection by the cuvette can often be accounted for. However, when working with thin wafers of powder samples, scattering and reflection can become more significant. One potential solution is to dilute the solid samples by adding a material with a similar refractive index, which minimizes scattering effects. In optical spectroscopy, transmittance (τ) is the ratio of transmitted (I_T) to incident light intensities (I_0). Its negative decadic logarithm which is generally denoted as absorbance (A).

$$A = -\log T = -\log I_T/I_0 \tag{3.5}$$

$$A = -\alpha\, d = -\varepsilon\, c\, d \tag{3.6}$$

There is an alternate definition for the natural logarithm. IUPAC recommends using "attenuance" instead of "absorbance," as attenuation can result from factors beyond absorption. In cases where absorption is weak, the relationship between absorbance and optical path length (d) can be expressed as (Lambert-Beer law) $A = \varepsilon c d$, where ε is the molar absorption coefficient, c is the concentration, and d is the path length (eq. (3.6)).

Figure 3.13 depicts the main components of a UV-vis spectrometer: the light source, monochromator, sample compartment (here, an integrating sphere for diffuse reflectance measurements), and detector. Gas discharge lamps used as radiation sources do not cover the entire spectral range. Diffuse reflectance spectra usually indicate the change from the deuterium lamp to the tungsten halogen lamp at around 320 nm. When using a scanning spectrometer, the beam passes through a rotating grating monochromator, scanning the energy across a desired range. Once the beam has traveled through the sample compartment, the detector measures and records the reflected intensity. Figure 3.13 illustrates a double-beam spectrometer simultaneously measuring signals from the sample and a reference. This is achieved using a chopper to alternate the beam's path between the sample and the reference. In contrast, single-beam spectrometers measure the spectra of the sample and the reference sequentially. The detector in this range is typically a photomultiplier tube.

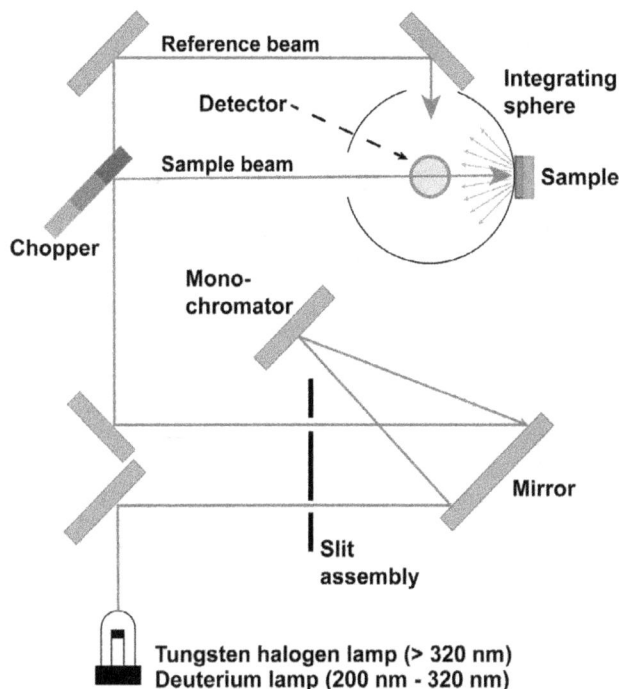

Figure 3.13: An integrating sphere with a UV-visible spectrometer serves as a tool for conducting measurements and obtaining diffuse reflectance spectra (reproduced from ref. [15]).

3.5.2 UV-vis spectroscopy in catalysis

UV-vis could monitor d-d transitions; it is possible to determine the oxidation state and coordination number of various transition metal cations. In the 1980s, Packet and Schoonheydt made significant breakthroughs in site-selective spectroscopic analysis on transition metal ions-exchanged zeolites, known as TMIs-zeolites. Figure 3.14a shows the emission spectra of Zeolite Y with various Mn^{2+} contents ($x > 0.4$) when excited with 413 nm light. As the Mn^{2+} content in Zeolite Y increases, the emission shifts from 650 to 685 nm. This shift is primarily caused by the interaction between the Mn^{2+} ions, with their delocalized d-electrons, and the multiple binding sites available for Mn^{2+} ions in the Zeolite Y structure [16]. As depicted in Figure 3.14b, a UV band was observed at around 48,000 cm^{-1}, indicative of isolated Cu^{2+} species in 2Cu/MCM-22 zeolite. Notably, no bands corresponding to CuO_x species at 22,500 and 40,000 cm^{-1} were detected, suggesting that most Cu species in 2Cu/MCM-22 are likely to exist in an isolated Cu^{2+} form. For xCu/MCM-22 zeolites with higher copper loadings (6–10 wt.%), the UV absorption bands at 40,000 and 22,500 cm^{-1} become more prominent. Notably, the band's intensity at 40,000 cm^{-1} surpasses that at 48,000 cm^{-1}, suggesting a higher concentration of CuO_x species in the

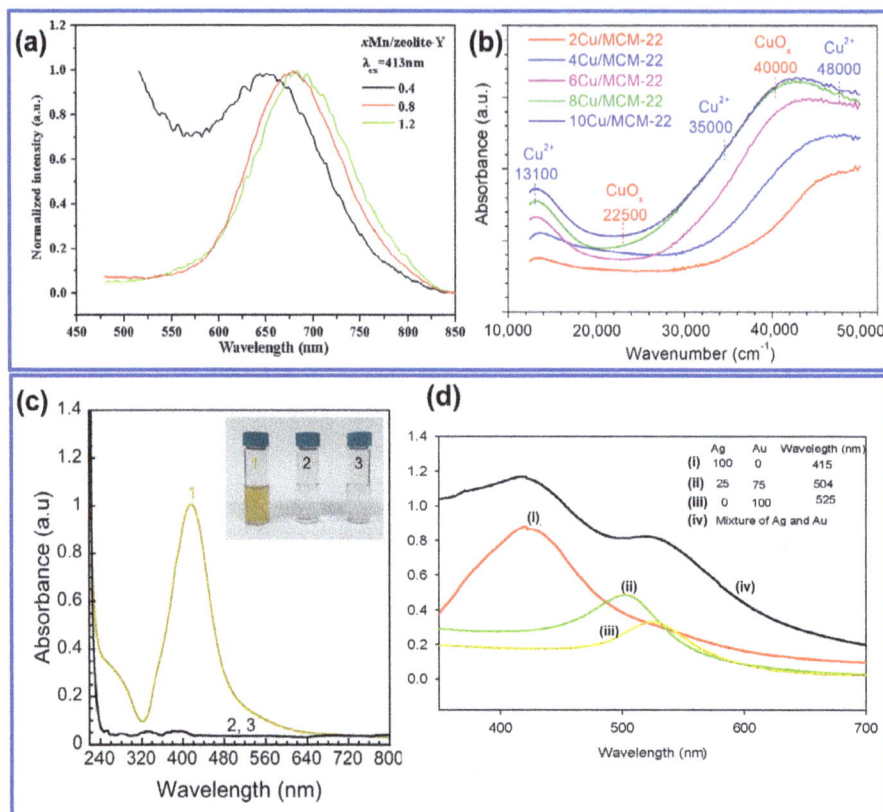

Figure 3.14: The emission UV-vis spectra of **(a)** xMn/Zeolite-Y (x = 0.4, 0.8, and 1.2) (adapted from ref. [14]); **(b)** xCu/MCM-22 zeolites with copper contents (adapted from ref. [15]); **(c)** UV-vis spectra of the AgNO₃ solution with sodium citrate (yellow curve-Vial 1) and without sodium citrate and light (black curve-Vial 2 and 3, respectively) (adapted from ref. [18]); and **(d)** UV-vis spectra for (i) Ag, (ii) Au, (iii) Ag-Au alloy NPs, and (iv) a mixture of pure Ag and Au NPs (adapted from ref. [19]).

zeolite. Based on the UV-vis analysis, isolated Cu^{2+} and CuO_x species concentration increases with increasing Cu loadings in xCu/MCM-22. Notably, for 2Cu/MCM-22, isolated Cu^{2+} species dominate, while for xCu/MCM-22 with high Cu loadings (>4 wt.%), CuO_x species become the majority [17]. UV-vis can be used to monitor the formation of metal nanoparticles (NPs) in real time. An example of this was synthesize Ag spherical NPs in the presence of sodium citrate as a reduction agent. The growth of Ag clusters was monitored by UV-vis spectroscopy. Initially, the NPs formed from an AgNO₃ solution in the presence of sodium citrate (Figure 3.14c, Vial 1) and without sodium citrate (Figure 3.14c, Vial 2). Both solutions were exposed to light at room temperature. In addition, the AgNO₃ and sodium citrate solutions were kept in the dark (Figure 3.14c, Vial 3). After 1 h of light exposure, the mixture of AgNO₃/citrate solution turned yellow. The UV-VIS spectrum of the yellow solution displayed a significant peak centered around 415 nm, indicative of the presence of Ag nanocrystals of ~40 nm diameter. The solution without sodium citrate, as well as the AgNO$_3$/citrate solution, kept in the dark, did not show any color changes, and their UV-VIS spectra remained unchanged. TEM imaging revealed the formation of various anisotropic Ag nanostructures, including triangles, rods, hexagons, and cubes, as well as nano prisms, depending on the presence of light excitation and citrate in the AgNO$_3$ solution. The experiments demonstrated that without both of these factors, the formation of such anisotropic nanostructures was not observed [18]. Also, Ag-Au bimetallic nanoparticles were chemically prepared by the reduction of AgNO$_3$ and HAuCl$_4$ using C$_6$H$_5$O$_7$Na$_3$ as the reducing agent as well as the capping agent and then immobilized on the surface of the oxidized polypyrrole/ a glassy carbon electrode (PPyox/GCE). Ag-Au nanoparticles were synthesized by adding 49 mL of water into a 100 mL round-bottomed flask. 2% (w/v) sodium citrate (0.5 mL) and 10 mM HAuCl$_4$ (0.5 mL) solution were added to the water, and the reaction mixture was heated to 92 °C. Subsequently, a mixture of 10 mM AgNO$_3$ (0.5 mL) was added, and the temperature was maintained between 90 °C and 92 °C while refluxing for 1 h. The Ag/Au ratio of 1:3, the volume of the mixture was adjusted by adding 0.125 mL of 10 mM AgNO$_3$ and 0.375 mL of 10 mM HAuCl$_4$, resulting in a final volume of 0.5 mL. During the reaction, a colour change was observed, with the solution turning dark red, indicating the formation of Ag-Au nanoparticles. The NPs were characterized by UV-visible spectroscopy technique (Figure 3.14d) which confirmed the homogeneous formation of the bimetallic alloy NPs in the range of 20–50 [19].

Furthermore, UV-vis spectroscopy can also be used to monitor the formation of characteristic species within the structure of zeolite-like materials during the methanol-to-olefins (MTO) reaction [20–24]. Figure 3.15a presents time-resolved UV-vis diffuse reflectance spectra for the MTO reaction during different stages. At lower temperatures (300 °C), absorption bands at around 30,000 and 24,500 cm^{-1} are identified as low methylated benzene and methylated naphthalene carbocations, respectively. At higher temperatures (400 °C), the catalyst deactivation process involves broadening the absorption band at 24,500 cm^{-1}, accompanied by the emergence of new bands at around 20,000 and 16,700 cm^{-1}. This suggests that at these temperatures, methylated naphthalene carbocations are not responsible for deactivation but rather neutral polyaromatic compounds and phenanthrene/anthracene carbocations hinder the diffusion of reaction products, contributing to catalyst deactivation in H-SAPO-3 [20]. In this additional study, as depicted in Figure 3.15b, when methanol flow is maintained at 300 °C, the primary species formed are highly methylated benzene carbocations (around 26,000 cm^{-1}) and monoenyl carbocations (around 33,650 cm^{-1}). However, when methanol flow is halted, the formation of olefins slows, and the UV-vis band shifts from around 26,000–25,000 cm^{-1}, suggesting the presence of methylated naphthalene carbocations. These species act as deactivating agents at lower reaction temperatures. When methanol flow is halted, and the reaction temperature reaches 350 °C, the formation of propylene and ethylene resumes, accompanied by a slight shift in the absorption band at around 25,000 cm^{-1} to lower wavenumbers, indicating demethylation of naphthalene carbocations. Additionally, typical UV-vis absorption bands associated with anthracene and

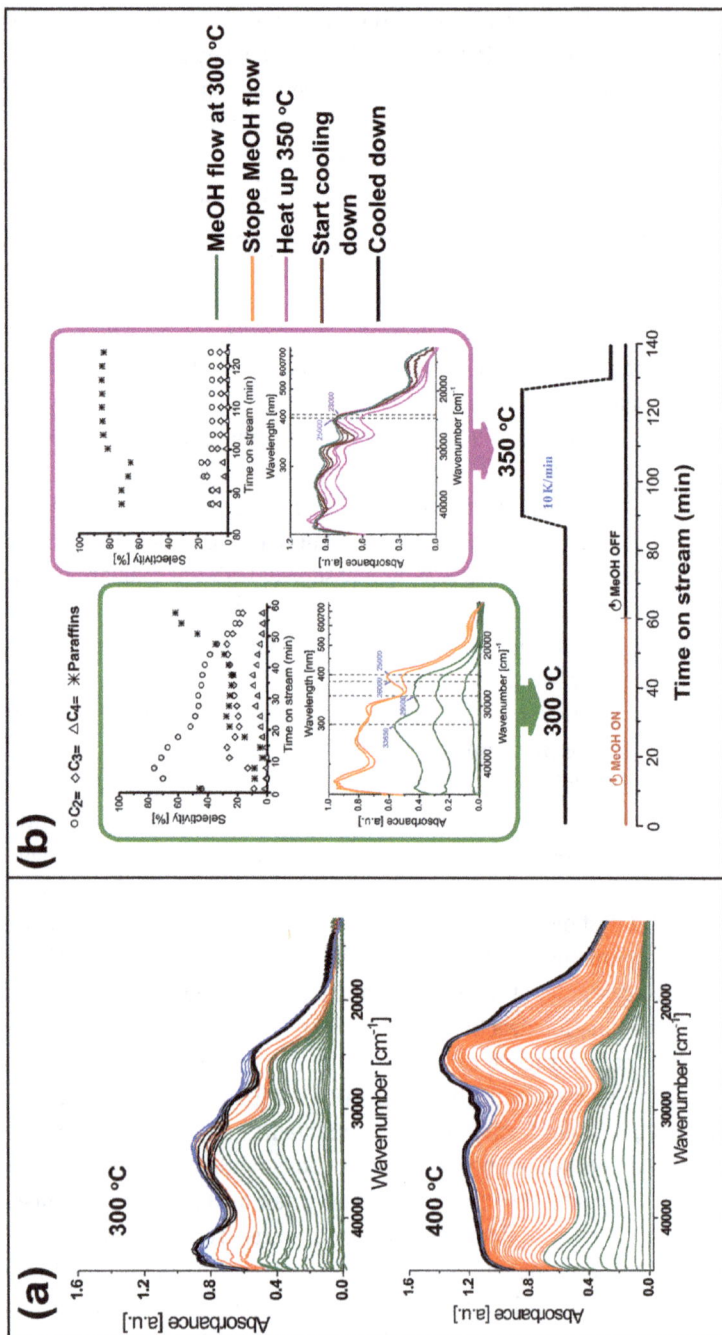

Figure 3.15: The time-resolved UV-vis diffuse reflectance spectra for the MTO reaction: **(a)** during different stages: Green for the induction period, red for full conversion, blue for the transition from 100% to 20% conversion, and black when the catalyst is deactivated; **(b)** the selectivity and operando UV-vis diffuse reflectance spectra of H-SAPO-34 were recorded at various stages as follows: at 300 °C during methanol flow (in green): at 300 °C while stopping methanol flow and flushing with He (in orange): upon heating up and maintaining the catalyst at 350 °C (in pink): During the cooling process back to room temperature (in brown): finally, at room temperature (in blue) (reproduced from ref. [20]).

phenanthrene carbocations are observed at 350 °C, signifying the formation of hydrocarbon deposits.

3.5.3 Diffuse reflectance UV-vis (DRUV-vis) spectroscopy

A technique used to measure the diffuse reflection of solid samples at different wavelengths to obtain information about their optical properties, such as absorption coefficient and scattering characteristics. Diffuse reflectance should be distinguished from the intensity resulting from specular reflection of the incident beam at the sample surface, which stems from only one interaction with the solid. On the other hand, their scattering properties influence the optical path length in samples or sample beds, which usually varies with the wavelength. Consequently, the absorption coefficient (denoted as K in DRS) is always related to a scattering coefficient, S, which accounts for the scattering behavior of the sample.

The relationship between the normalized absorption coefficient (K/S) and the reflectance measured at infinite sample thickness ($R\infty$), as derived from a phenomenological model that assumes diffuse illumination, weak absorption, and scattering as the primary mode of photon direction, is expressed as

$$\frac{K}{S} = \frac{(1-R_\infty)^2}{2R_\infty} = F(R_\infty) \tag{3.7}$$

$F(R\infty)$ is referred to as the Kubelka-Munk function, named after the model's authors. K and S, which determine the shape of the function, have units of reciprocal length. They can be individually determined through additional measurements, which is seldom done in catalysis. Typically, spectra are presented as $F(R\infty)$ versus wavelength (λ), or frequency ($\bar{v}$) [15].

In ultraviolet-visible (UV-vis) DRS spectroscopy, semiconductor band gaps are commonly determined, either direct or indirect. The DR UV-vis spectra for synthesized titanium dioxide (TiO$_2$) samples treated at 500 °C are displayed in Figure 3.16a. The reflectance peak maximum varies for the TiO$_2$ samples synthesized using different acids, such as sulfuric (sul), nitric (nit), and acetic (act) acids. TiO$_2$-sul displayed a peak maximum at 372 nm, shifting to 383 nm in TiO$_2$-ace and 402 nm in TiO$_2$-nit [25].

Figure 3.16b presents the DRUV-vis spectra of graphene oxide (rGO)-Zirconium dioxide (ZrO$_2$) nanocomposite samples along with bare rGO and ZrO$_2$. The bare rGO showed a minor band at 220 nm and a broader band at 305 nm, which can be attributed to $\pi \rightarrow \pi^*$ transitions of the aromatic C–C and C=O bonds, respectively. When rGO was incorporated into ZrO$_2$, the band's wavelength maximum (λ_{max}) corresponding to the O$_2 \rightarrow$ Zr^{4+} charge transfer transition shifted to lower frequencies. The rGO-ZrO$_2$ sample had a small band at 225 nm and a broad band centered at 325 nm. Studies have shown that the 250–350 nm absorptions observed in ZrO$_2$ nanoparticles can be attributed to O$_2 \rightarrow$Zr^{4+} charge transfer transitions [26]. The increase in rGO loading led

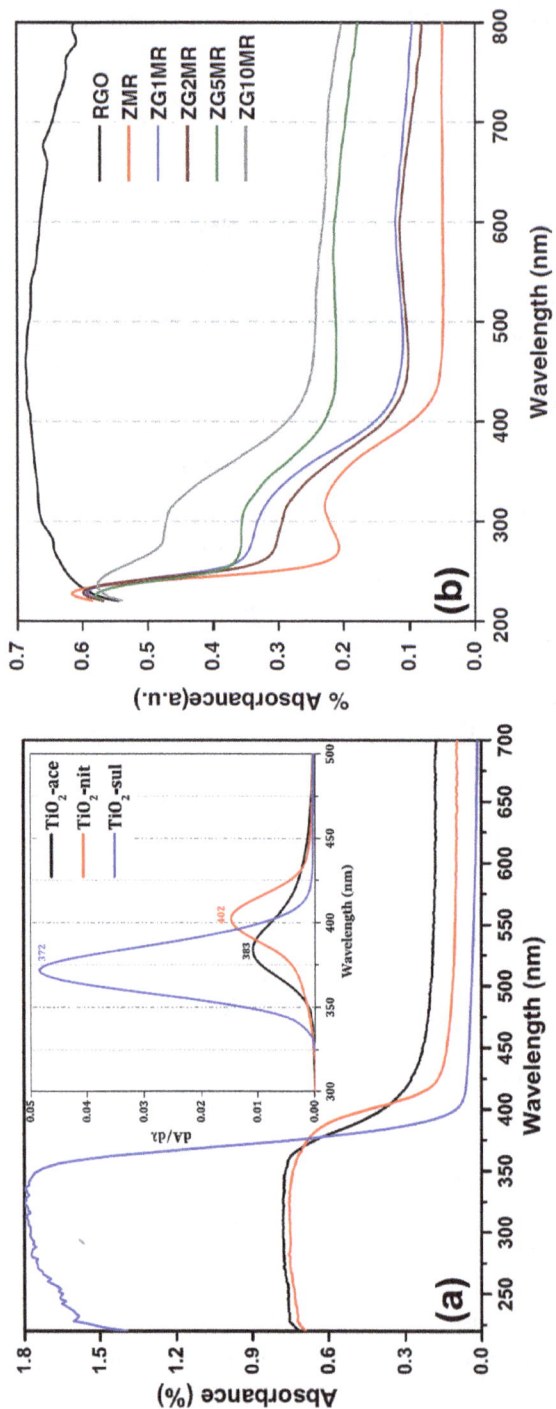

Figure 3.16: Diffuse reflectance ultraviolet-visible (DR UV-vis) spectra for **(a)** calcined titanium dioxide (TiO$_2$) samples (adapted from ref. [25]) and **(b)** the calcined rGO-ZrO$_2$ nanocomposite samples (adapted from ref. [26]).

to a further shift in the peak and a broadening of the peak, likely due to the π-conjugation network formed within the rGO-ZrO$_2$ nanocomposite.

The Tauc plot is a method used to determine the band gap energy of semiconductors from their absorption spectra. It is based on the relationship between the photon energy (hv) and the absorption coefficient (a) of a semiconductor, which is described by Tauc's law:

$$(a\ hv)^2 = A\ (hv - E_g)^m \tag{3.8}$$

where E_g is the band gap energy, A is a constant, and m is an optical parameter that depends on the absorption mechanism. By plotting $(ahv)^2$ against Energy (eV) and extrapolating to the hv axis, the band gap energy can be determined. For instance, back to the TiO$_2$ synthesized using different acids. TiO$_2$ exists in two primary forms: anatase and rutile. The bandgap energies for anatase and rutile forms of TiO$_2$ have been reported to be 3.2 eV (380 nm) and 3.0 eV (415 nm), respectively. The bandgap energies were determined for the calcined samples using the Tauc plot. The data showed that the bandgap energy for TiO$_2$-sul was higher at 3.12 eV, while it was lower for TiO$_2$-ace (2.99 eV) and TiO$_2$-nit (2.97 eV). TiO$_2$ bandgap decreased when the rutile phase became more predominant in the sample. Studies have indicated that the valence band of both anatase and rutile phases consists mainly of O 2p states, whereas the conduction band predominantly comprises Ti 3d states [25]. Another example, the DR UV-vis spectra were recorded in the 200–800 nm range (Figure 3.17a). The synthesized analcime (ANA) zeolites displayed a primary absorption band in the UV region at 250 nm, associated with the charge transfer process inherent in the zeolite structure that synthesized via various precursor: aluminum nitrate (ANA-nit), aluminum sulfate (ANA-sul), aluminum isopropoxide (ANA-isop), sodium aluminate (ANA-sodalu) and aluminum chloride (ANA-chl). The zeolite framework possesses electron donor and acceptor properties, contributing to this absorption band. Interestingly, the diffuse reflectance spectra of ANA-nit, ANA-isop, and ANA-chl samples reveal an additional low-intensity peak in the UV region at 378 nm, which can be attributed to the ligand-to-metal

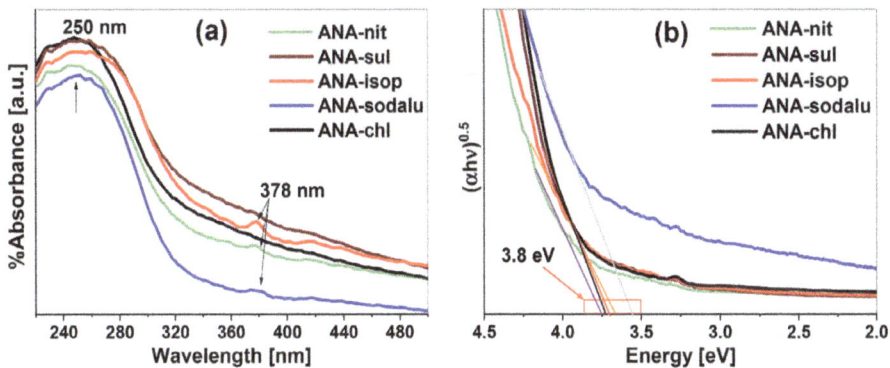

Figure 3.17: (a) Diffuse reflectance ultraviolet-visible (DR UV-vis) and **(b)** the Tauc plots for synthesized Analcime samples (adapted from ref. [27]).

charge transfer phenomenon (specifically, $O^{2-} \rightarrow M^{3+}$). Figure 3.17b shows Tauc plots of the synthesized Analcime samples with a calculated bandgap energy of ~3.8 eV [27].

3.6 Nuclear magnetic resonance spectroscopy

Nuclear magnetic resonance (NMR) is a powerful analytical technique that is used to study the structure and properties of both solid and liquid samples. It relies on the interaction between nuclear spins and an applied magnetic field to provide detailed information about the molecular environment and structure of a sample. There are several different types of NMR techniques, each with its own unique applications and capabilities. These include solution-state NMR, which is used to study molecules in solution, and solid-state NMR, which is used to study molecules in solid form. NMR has numerous potential applications in a wide range of fields including chemistry, biology, material science, and medicine. Some potential uses of NMR include (i) structure determination of molecules and materials, (ii) monitoring chemical reactions in real time, (iii) characterization of the structure and dynamics of biological macromolecules, (iv) analysis of fluids and solid materials, (v) investigation of catalysis and reactions in porous materials, (vi) determination of diffusion coefficients and transport properties, (vii) identification of unknown compounds and mixtures, (viii) quality control and process monitoring in pharmaceutical and chemical industries, etc.

3.6.1 NMR concept

NMR is rooted in the nuclear property of having intrinsic angular momentum (spin). This angular momentum is also known as "spin angular momentum" or "spin," it is distinct from rotation-based angular momentum. Unlike angular momentum attributed to rotation, spin is an inherent characteristic of a particle, not contingent upon its rotation. The energy associated with the nuclear spin in a magnetic field is influenced by the gamma (Y) and the strength of the magnetic field (B_0). For a spin of 1/2, there are two possible energy levels corresponding to two orientations, represented by $m_I = +1/2$ and $-1/2$ (see Figure 3.18). When a weakly oscillating magnetic field of en-

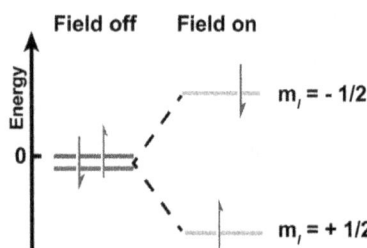

Figure 3.18: Energy levels for a nucleus with spin quantum number ½.

ergy hv is applied to the system and the energy of this field corresponds to the separation between the energy levels in the magnetic field, a signal can be observed from the resonating nucleus.

3.6.2 Solid-state NMR in catalysis

Characterizing active sites on the surface of catalysts is often a crucial objective for studies in this field. This includes identifying acidic or basic sites, as well as determining the structure and sometimes the oxidation state of active species present on the surface of catalysts. One advantage of NMR is that it is particularly sensitive to the short- to medium-range order of a material's structure, making it well-suited to studying amorphous or weakly crystalline materials. Amorphous materials lack a well-defined crystal structure and can be difficult to characterize using other techniques, such as X-ray diffraction (XRD). Still, NMR can provide valuable information about their local structure and composition. A variety of NMR methods are used to characterize solid materials. Nuclei whose magnetic properties make them easy or difficult to detect in NMR are categorized based on their magnetogyric ratio and the natural content of their magnetic isotope. There are 13 easily detectable nuclei by NMR: ^1H, ^7Li, ^{11}B, ^{14}N, ^{19}F, ^{23}Na, ^{27}Al, ^{29}Si, ^{31}P, ^{51}V, ^{129}Xe, ^{133}Cs, ^{195}Pt [28]. Several classes of solids are of particular interest in catalysis, such as silicon- and aluminum-containing materials. Some examples of these materials include zeolites, metal-organic frameworks (MOFs), and silica-based materials. Silica – and alumina-based materials are among the most used materials in catalysis, as they can either support the catalytically active component or act as the catalyst themselves. The majority of solid-state NMRs used are ^{29}Si and ^{27}Al.

In Silica materials, the characterization of the bulk structure can be determined from the analysis of the Si-O-Si bond, which can be correlated with the chemical shift observed in the ^{29}Si MAS NMR spectrum. While the structure of porous silicas frequently used ^{29}Si and ^1H NMR methods. The methods encompass ^1H Magic Angle Spinning (MAS), ^1H Cross-Polarization, and Magic Angle Spinning (CRPMAS), as well as ^{29}Si-^1H Constant-Time HETORMAS (CP-MAS). These methods are employed to characterize the surface of silicas in terms of the two major types of silanol species: Q_3 [(-O)-$_3$SiOH] and Q_2 [(-O)$_2$Si(OH)$_2$] species. Also, distinguish hydrogen-bonded silanols from non-hydrogen-bonded silanols [29].

Zeolite materials, ^{29}Si MAS NMR spectra give information about the nearest neighboring of Al atoms to Si atoms in the zeolite framework. It was reported that the ^{29}Si MAS NMR resonances above –110 ppm could be assigned to Si(4Si) or Si(0Al) sites, that is, Si-atoms with no Al neighbors environment. The resonances between –103 and –108 ppm could be assigned to Si(3Si,1Al) sites, that is, Si atoms with one neighboring Al atom. The resonances with chemical shifts below – 100 ppm representing Si (2Si,2Al) sites with two Al neighbors (see Figure 3.19a and b).

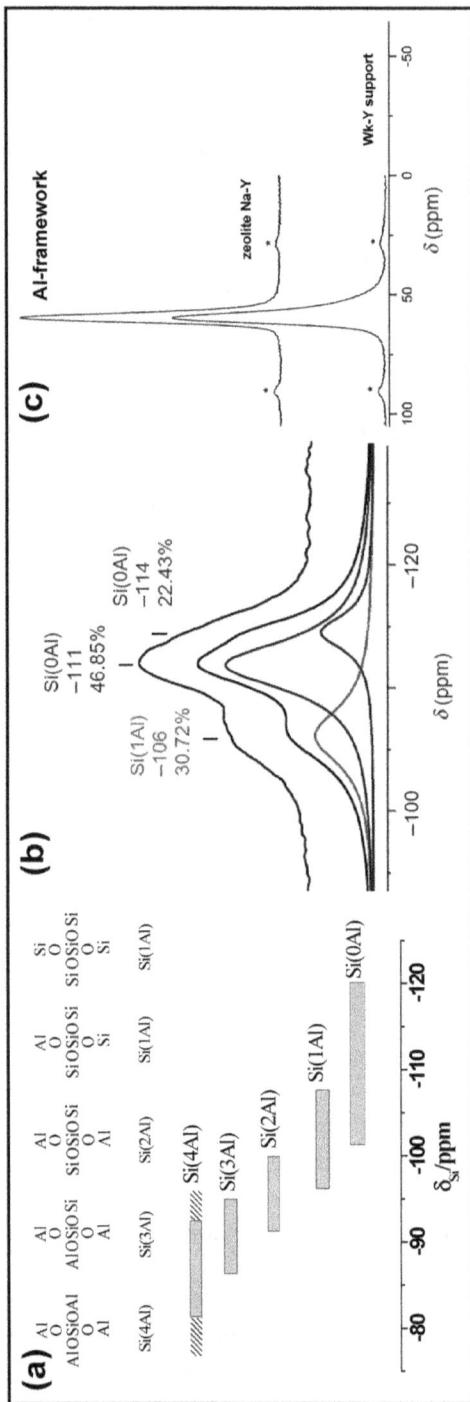

Figure 3.19: (a) ^{29}Si chemical shift of Si(nAl) units in zeolite framework depends on the second-neighbor environment of silicon, **(b)** examples of ^{29}Si MAS NMR spectra for H-ZSM-5 (Si/Al = 13) (adapted from ref. [28]), and **(c)** zeolite Na-Y and Wk-Y structure (adapted from ref. [30]).

The ^{27}Al MAS NMR spectra of aluminosilicates are predicted to be simpler than the ^{29}Si MAS NMR spectra. According to Loewenstein's rule, only a single peak in the NMR spectrum should be observed, corresponding to alumina atoms in a tetrahedral environment, for example, zeolite Y (Figure 3.19c). However, six- and five-coordinate alumina

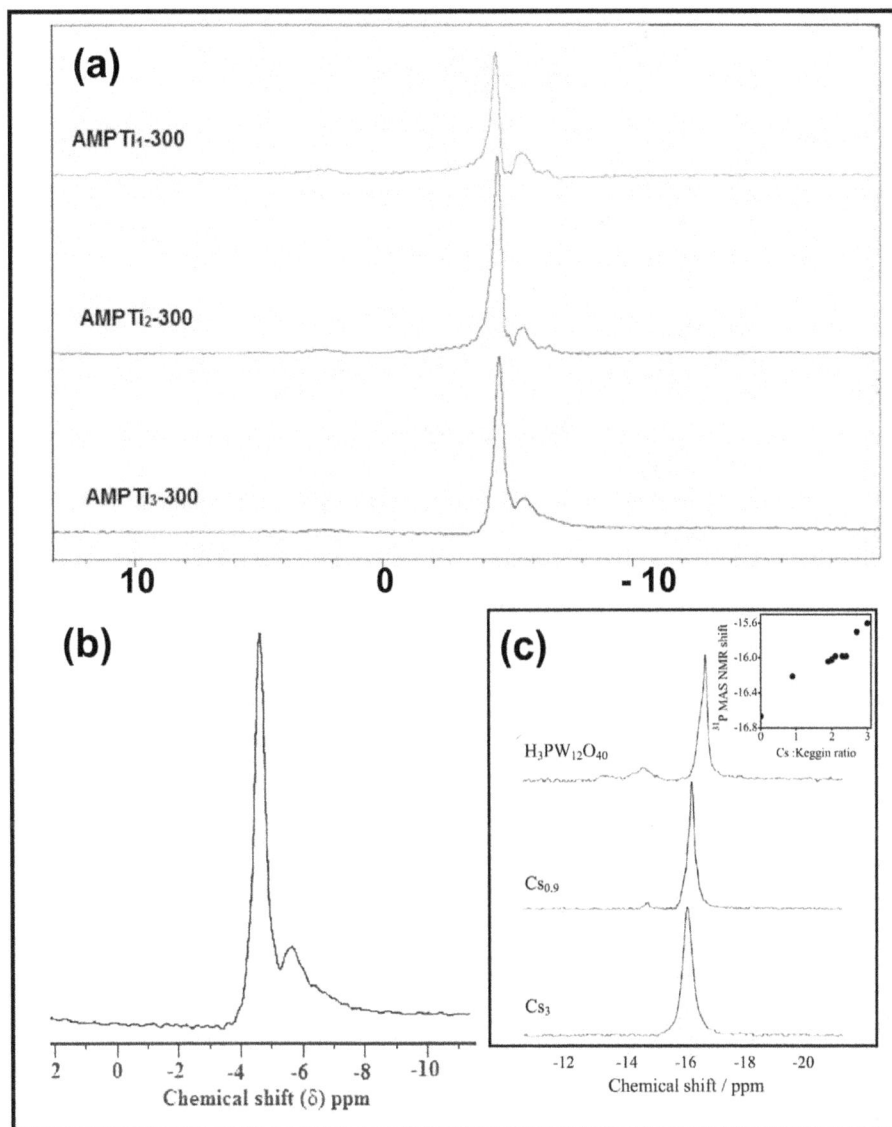

Figure 3.20: The ^{31}P MAS NMR chemical shift for **(a)** titanium is supported on AMP (adapted from ref. [27]), **(b)** 12-molybdophosphoric acid (AMPA) (adapted from ref. [32]), and **(c)** Cs are loaded to HPW (adapted from ref. [33]).

also exist, adding complexity to the NMR spectrum, known as extra-framework in zeolites. Zeolites have a negatively charged framework due to the presence of tetrahedrally coordinated aluminum atoms, which require positive charge compensation. Either tetrahedrally coordinated protons in hydroxyl bridging groups or exchangeable extra-framework cations can offset this charge.

Phosphorus-based materials include phosphates, phosphonates, heteroplyacids (HPAs), ammonium salt of 12-molybdophosphoric acid, etc. ^{31}P MAS NMR is useful for investigating phosphorus-containing materials and provides important structural, chemical, and dynamic information about the phosphorus atoms in these materials. For example, Figure 3.20a in the phosphorous ^{31}P NMR spectrum of Titanium supported on ammonium salt of 12 molybdophosphoric acid (AMP). This spectrum was obtained after calcination at a temperature of 400 °C. The spectrum shows a major peak at –8.2 ppm, attributed to Titanium species. Additionally, there is a peak at 0 ppm and a broad hump at –4.5 ppm. The peak at 0 ppm is believed to result from phosphorus oxides formed due to the decomposition of the Keggin ion. In comparison, the peak at –8.2 ppm can be associated with lacunary species [31]. Figure 3.20b represents the 12-molybdophosphoric acid (AMPA), the chemical shift of ^{31}P NMR found at – 3.9 ppm for $H_3PMo_{12}O_{40} \cdot 30H_2O$ and –2.9 ppm for $H_3PMo_{12}O_{40}$ [32]. And Figure 3.20c, the ^{31}P NMR spectra of heteropolytungstate (HPW), Cs1, and Cs3 samples dried at 100 °C, there is a single peak that shifts from $\delta =$ –16.7 ppm to –15.6 ppm with the substitution of all three protons by cesium (Cs). A small peak at –14.6 ppm was observed in the parent and Cs1 samples, attributed to HPW with less water content. The ^{31}P NMR chemical shift changes are correlated with the water content, as the hydration level decreases with an increase in Cs content in the secondary structure of the polyoxometalate [33].

3.7 X-ray diffraction technique

XRD has become a standard laboratory technique for studying condensed crystalline matter. As a non-destructive method, it provides details on crystal structure, lattice constants, and particle size of synthesized materials [34]. This technique is commonly employed for the identification of unknown crystalline materials. When applied under in situ conditions, XRD can unveil crucial processes during catalyst preparation, pretreatment, activation, or deactivation stages. This technique shines as a valuable tool for investigating size modifications in response to dopant changes, temperature, and synthesis time, among other parameters. It also serves as a quality control mechanism during catalyst fabrication and a means to analyze specific properties such as domain size, crystallinity or disorder, composition of mixed phases, texture, and solid-state reactions involving crystalline phases. By conducting an XRD experiment, one can identify whether a sample is at least partially crystalline and recognize the presence of one or multiple crystalline phases. Further evaluation enables determination of the relative quantities of crystalline components. Additionally, the crystal structures can be refined, and some microstructural parameters such as crystallite size and lattice strain can be extracted by analyzing the line profile. Crystalline structures involve the organization of ions, atoms, or molecules in a three-dimensional, ordered manner, defined by repeating unit cells that establish inter-atomic bond distances and angles. Solid material structures can take the form of single crystals, polycrystalline material, or amorphous solids, as discussed in Chapter 1. X-ray crystallography has been widely applied to various inorganic, organic, and biological materials that form crystals. Numerous X-ray techniques have been developed for analysis such as single crystal diffraction (SCD), X-ray powder diffraction (XRPD), small-angle X-ray scattering, grazing incidence angle diffraction, back-reflection Laue diffraction, and X-ray reflectivity. SCD is commonly used for crystalline materials, while XRPD is typically applied to polycrystalline and amorphous substances.

3.7.1 X-ray diffraction concept

The experiments conducted by Wilhelm Röntgen, known as cathode ray tubes (CRT), served as a means to generate X-ray radiation, as shown in Figure 3.21a. X-ray monochromatic radiation is created by heating the filament in the cathode, which releases electrons and subsequent bombardment with metal (such as copper (Cu), molybdenum (Mo), cobalt (Co), silver (Ag), or tungsten (W)) results in the emission of characteristic monochromatic photons. When the paths of such sources overlap, they can interfere constructively, destructively, or partially destructively (Figure 3.21b).

Bragg's law delineates the prerequisites for diffraction in a single crystal. It serves to calculate the interspace and distance between layers of atoms, where it takes into account the path difference between the two rays scattered at the first and second layers. As X-ray photons with precise angles of incidence arrive at the regular interplanar spacing (d), they generate a constructive interference (in-phase) of wavelength (λ) and form an angle of incidence and an angle of reflection equal to Θ. This sets up the geometrical formation presented in the inset of Figure 3.21b, which enables calculation of the extra distance traversed by the bottom ray. The path lengths of the two beams are represented as (ab) and (bc) in terms of wavelengths when diffraction occurs. Then the extra path distance followed by the bottom ray must be equivalent to an integer multiple of λ, represented as (n) λ:

$$n\lambda = ab + bc \tag{3.9}$$

When the geometry is rewritten using the **Pythagorean formula** of ($a^2 + b^2 = c^2$), where a is the side of a right triangle, b is the side of the left triangle, and c is the hypotenuse (eq. (3.9)), we obtain the following equation:

$$d\sin\theta_o = ab \text{ and } d\sin\theta = bc \tag{3.10}$$

$$n\lambda = d\sin\theta_o + d\sin\theta \tag{3.11}$$

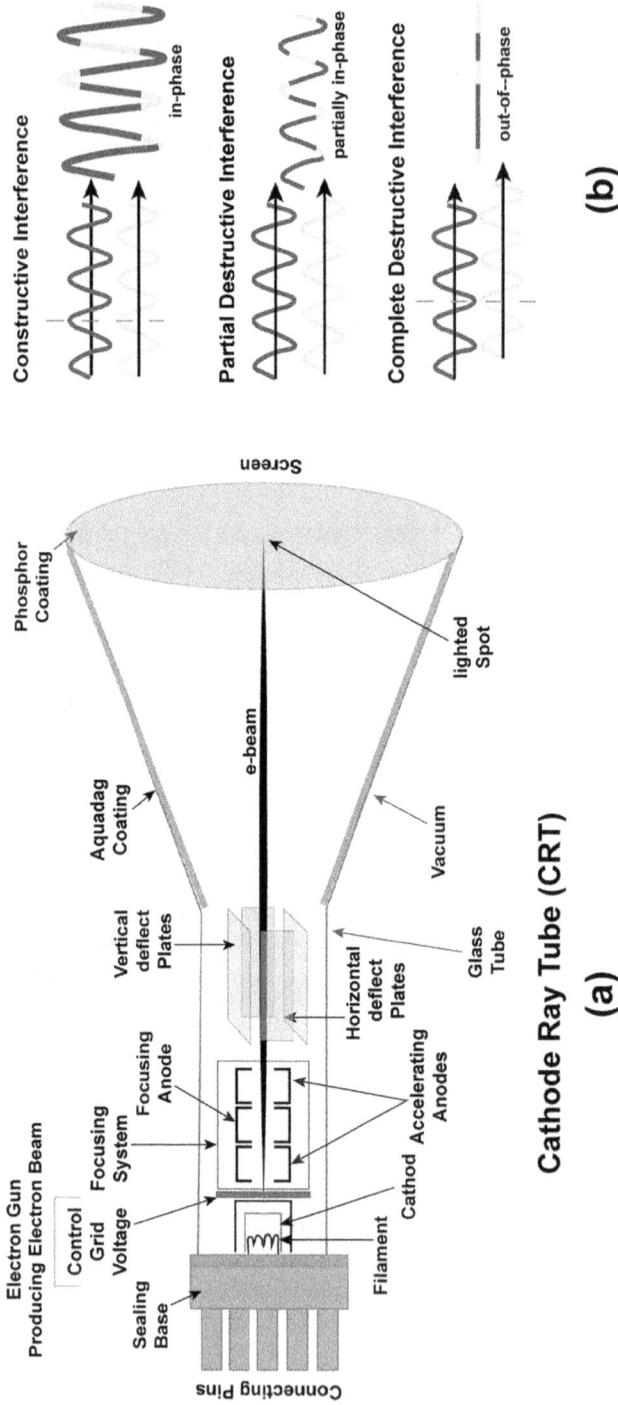

Figure 3.21: (a) A typical internal structure of a cathode ray tube used in a cathode ray oscilloscope. **(b)** The different types of wave interference.

The incident radiation and the lattice plane must adhere to the Bragg condition or Bragg's law (eq. (3.12)) for the diffraction of the incident radiation across successive lattice planes, which lead to the emergence of measurable intensity peaks within the diffracted beam.

$$n\lambda = 2d \sin \theta_{hkl} \tag{3.12}$$

where n is the diffraction order of reflection, λ represents the X-ray wavelength, d is the characteristic spacing between interatomic layers, and θ_{hkl} indicates the angle between the incident X-ray beam and the scattering plane corresponding to the crystal planes defined by hkl. In the illustration provided in Figure 3.22a, it suffices to consider only two layers of atoms to explain the constructive interference if the Bragg equation is fulfilled.

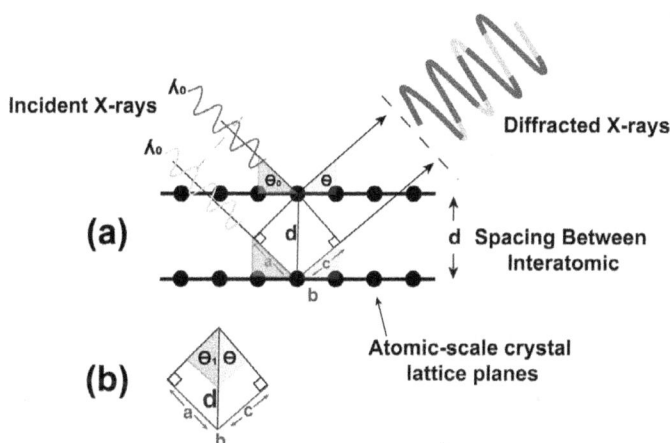

Figure 3.22: (a) Schematic illustration of Bragg's law: two layers of atoms with interlayer spacing *d* and idealized amplitude of the scattered waves. (b) The geometry formed from the incident and diffracted waves of angle *θ*.

In the realm of catalysis, XRD is consistently applied to polycrystalline powders, wherein the crystallites are oriented in a random manner across all conceivable directions. When a powdered sample is positioned within an X-ray or neutron beam, all possible inter-atomic lattice planes (*hkl*) will undergo diffraction at the angle (θ_{hkl}) stipulated by the Bragg eq. (3.12). This condition is met for the *hkl* plane in any orientation accessible through the rotation of θ_{hkl} around the primary beam, taking into account all possible scattering angles. All these orientations can be in a powder sample. The diffraction pattern takes the form of a funnel generated by the rotation of the deflection angle ($2\theta_{hkl}$) around the primary beam. These coaxial, non-uniformly spaced cones (called Debye-Scherrer cones) are uniquely associated with a particular set of lattice planes. When these cones are projected onto a flat surface normal to the incident beam, they produce a series of concentric circles known as Bragg-Debye rings or Debye-Scherrer rings [35, 36], as depicted in Figure 3.23.

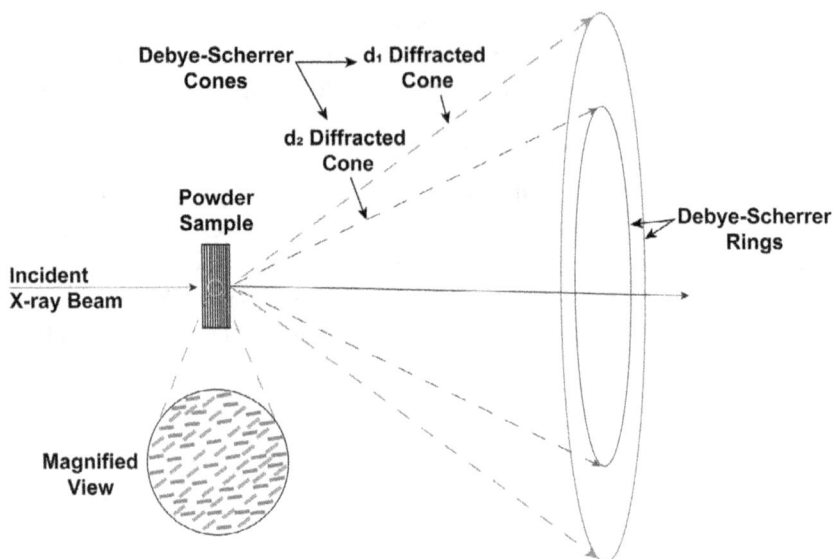

Figure 3.23: Production of diffraction cones from a powder sample illuminated by an incoming X-ray beam.

A diffractometer consists of an X-ray source combined with (i) beam optics, (ii) a sample holder, (iii) a detector, and (iv) a device capable of scanning the detector across a broad range of angles relative to the incident beam. In powder diffraction studies, most investigations are conducted with CuKα radiation, which offers a favorable balance of intensity and resolution, along with advantages in the production and maintenance of the X-ray tube. For elements susceptible to fluorescing under CuKα radiation, alternatives like CoKα or MoKα may be preferable. When employing characteristic CuKα radiation (λ = 1.542 Å), extraneous Cu lines are filtered out using metal filters. The width of the incident radiation can be further reduced by a monochromator crystal set to $CuK\alpha_1$ at 1.542 Å [15]. As illustrated in Figure 3.23, diffraction signals can be recorded by rotating the detector around the sample in a plane, thereby truncating the diffraction cones in half.

The Bragg-Brentano geometry (θ–2θ geometry), as depicted in Figure 3.24, operates via reflection. A flat and rotatable sample surface is used for placing the sample, and the detector is moved around it by a goniometer rotating twice the angular speed. The setup utilizes the para-focusing principle, which dictates that the divergence of the radiation arriving at a flat sample from a point source is reversed in the diffracted beams if the detector is positioned on a circle defined by the source and an osculation point on the sample. As the detector moves on the goniometer circle (depicted in black in Figure 3.24), the focusing circle also changes, as evidenced by the differences between the two measurement positions. The θ–2θ geometry is the most commonly employed in catalysis research. The measurements can take anything from minutes to days, depending on the characteristics of the sample. XRD is employed to determine crucial information about

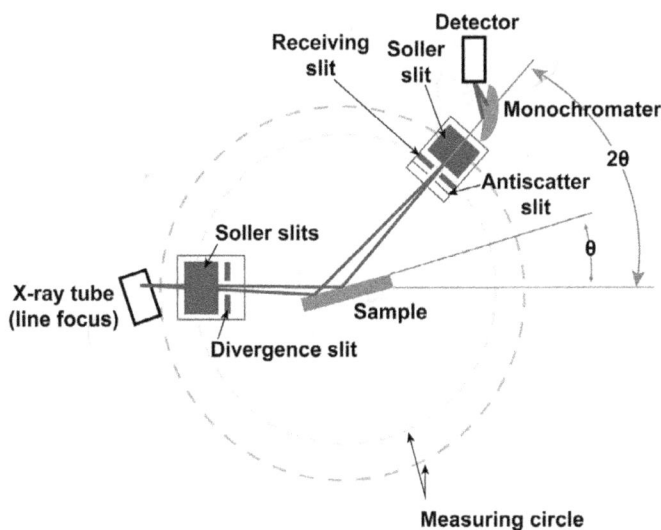

Figure 3.24: The Bragg-Brentano geometric arrangement of X-ray diffractometer.

the crystalline structure, degree of crystallinity, and particle size of synthesized materials, as well as to study modifications in materials as a function of changes in dopants and deactivation in spent catalysts.

3.7.2 X-ray diffraction analysis in catalysis

In most diffractograms, examples are provided based on Cu$K\alpha$ radiation. A powder XRD pattern serves as a unique fingerprint for a given material's structure (see Figure 3.25a and b), as illustrated by the identification of distinct cubic structures for different gold (Au) and silver (Ag) of pure metals [37, 38]. Another example demonstrates the same compound, ZrO_2 powders with varying structures (Figure 3.25c) in monoclinic (m-ZrO_2), tetragonal (t-ZrO_2), and cubic (c-ZrO_2) structures [39].

The interpretation of diffractograms commences with the process of indexing reflections, which is conveniently accomplished through the comparison of experimental patterns with reference patterns stored in databases like the **Powder Diffraction File** maintained by the **International Center for Diffraction Data**. Such comparisons are typically facilitated by software packages provided with modern diffractometers. In Figure 3.25c, the initial pattern aligns well with the reference pattern for the cubic (c-ZrO_2) structure. The peaks at 2θ = 30.3°, 35.14°, 50.48°, and 60.2° in cubic ZrO_2 according to JCPDS CAS number 27-0997 [39]. Notably, (i) there are no additional reflections in the diffractogram, and peak intensities may vary, as observed in Figure 3.25c, where the cubic c-ZrO_2 reference pattern displays an intense diffraction peak at 2θ = 50.48°, a difference from the experimental result. (ii) There might be discrepancies in peak positions

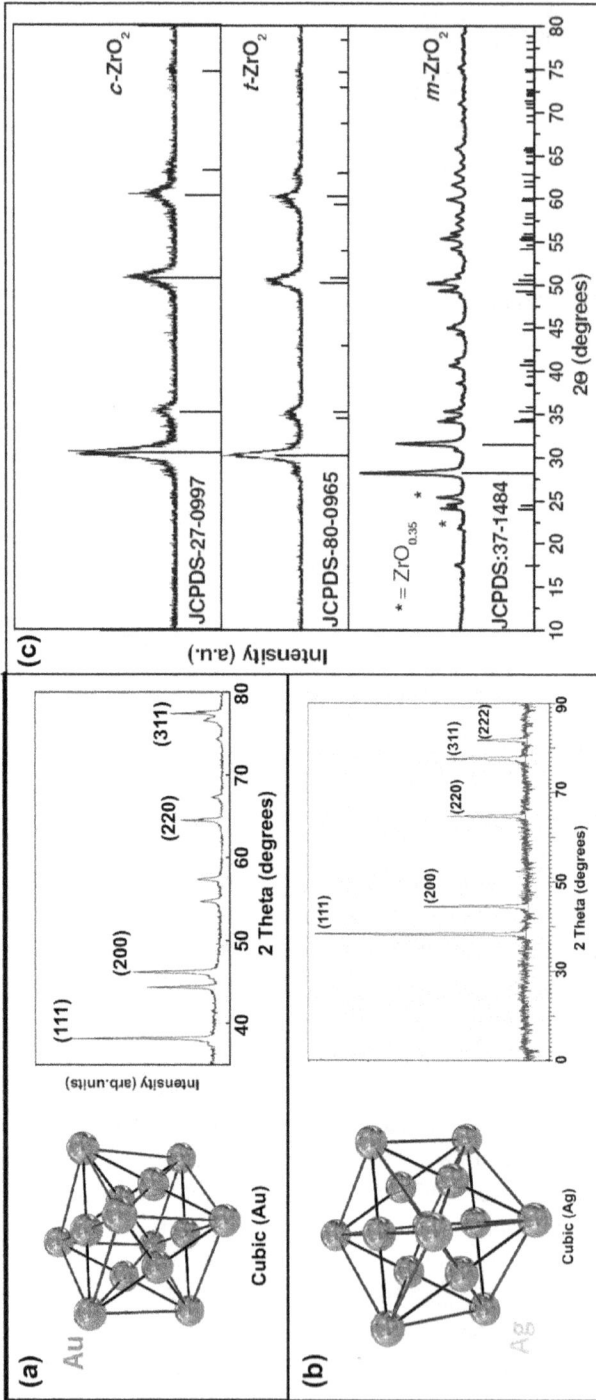

Figure 3.25: X-ray diffraction (XRD) pattern of cubic: **(a)** gold (Au), XRD (reproduced from ref. [37]), **(b)** silver (Au), XRD (reproduced from Ref. [38]), and **(c)** different structure of ZrO₂ XRD (adapted from ref. [39]).

between the experimental and reference patterns, as reflected in the tetragonal phase (t-ZrO_2) sample from pure ZrO_2 [JCPDS 80-0,965], demonstrating peaks at $2\theta = 30.2°$, $35.2°$, $50.6°$, and $60.2°$. (iii) Additional peaks might be present, indicating the existence of extra crystalline components. The m-ZrO_2 sample exhibits intense diffraction patterns at $2\theta = 24.2°$, $28.2°$, $31.4°$, and $34.3°$, corresponding to the monoclinic ZrO_2 crystal phase [JCPDS 37-1484]. Furthermore, the observation includes a prominent peak at $2\theta = 25.4°$, alongside a minor peak at $2\theta = 22°$, that are not indexed to the monoclinic ZrO_2 phase reference. These peaks can be indexed to oxygen-deficient zirconium oxide, the $ZrO_{0.35}$ phase [JCPDS: 17-0,385, hexagonal, space group P6322], observed in catalytic materials [39]. Due to small particle sizes, strain, or structural disorders, reflections are frequently broad, and intensity should be assessed based on peak areas rather than peak heights. Extremely broadened line shapes, commonly observed in catalytic materials, can lead to a decrease in the intensity background. Platelet-like or needle-like particle morphologies can induce a preferential crystalline orientation in the sample. Moreover, XRD provides the opportunity to delve into more data, such as comparing the intensity of reflections in the diffractogram to that obtained from a well-crystallized sample of the same phase, in order to assess the degree of crystallinity. The XRD analysis applied to synthesized materials is based on the principle that intensities are, in principle, proportional to the amounts of the corresponding phase present. However, X-ray amorphous material is not detectable through this procedure. Therefore, analysis via comparison with reference of object materials is the better option. Examples include the ZSM-5 structure synthesized using different methods compared with ZSM-5 reference [40] (Figure 3.26a) and synthesized using different Si/Al ratios [21] (Figure 3.26b).

XRD can also identify reference materials, such as carbon-supported Ag, $Au_{25}Ag_{75}$, $Au_{50}Ag_{50}$, $Au_{75}Ag_{25}$, and Au nanoparticle catalysts, mixed with sample material [41] (Figure 3.26c). By analyzing the diffractograms, the scattering efficiencies of the mixture components can be determined, allowing for access to their amounts in unknown mixtures based on the diffractograms measured with the reference admixed. X-ray analysis of amorphous material may lead to an underestimation of the sample weight. Additionally, there are other types of materials that are undetected by XRD or hinder the detection of specific components. As particle size decreases, the peaks in XRD patterns may become broader and ultimately disappear, especially for particles smaller than 2 nm in size, particularly for metals and materials with large unit cell structures, like zeolites. If particle sizes are distributed, the lower end of the distribution may fail to contribute to signal intensity, meaning that the reflection better represents the higher end. XRD also enables tracking of structural deactivation in the studied catalyst, for example, H-Ferrierite zeolite catalyst (Figure 3.27). Important differences exist between the fresh and deactivated samples, including a decrease in the first reflection's intensity (Figure 27b) and a shift toward lower angles (Figure 3.27c) [42].

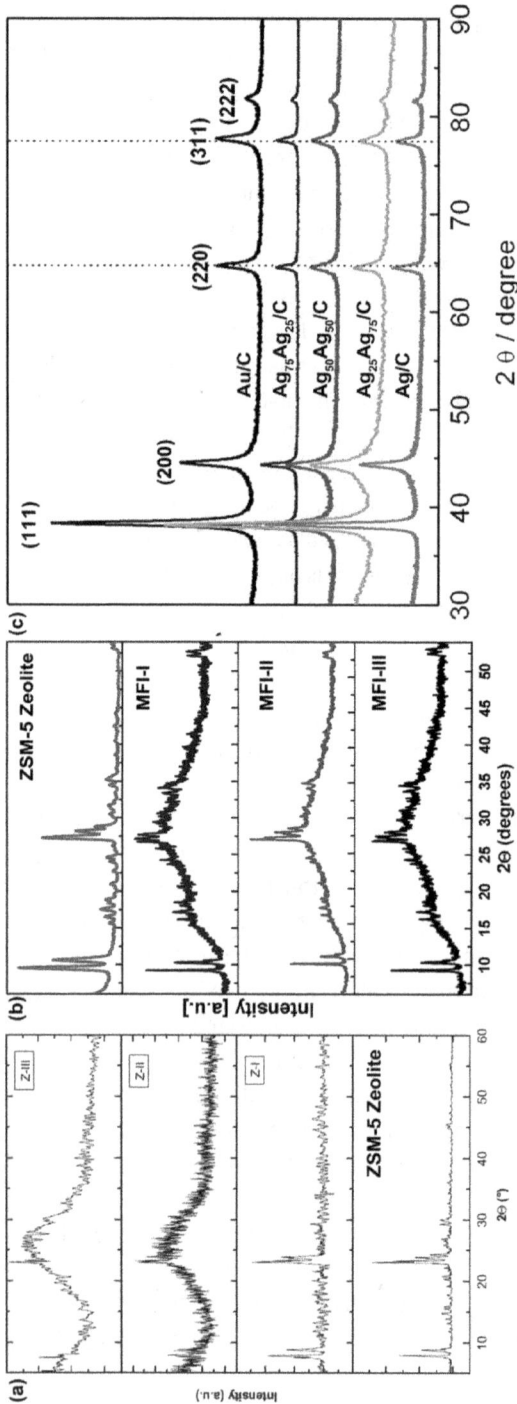

Figure 3.26: X-ray diffraction (XRD) pattern of ZSM-5: **(a)** compared the synthesized materials with ZSM-5 structure as a reference, Z-I: synthesized with seed with hydrothermal at 190 °C for 9 days, Z-II: synthesized with seed with hydrothermal at 120 °C for 2 days, Z-III: synthesized without seed using two temperatures at 180 °C and 120 °C. The XRD is adapted from Ref. [40], **(b)** compared the ZSM-5 structure as a reference with synthesized materials, MFI-I: Si/Al = 50, MFI-II: Si/Al = 45, and MFI-III: Si/Al = 40. The XRD is adapted from Ref. [21], and **(c)** compared the carbon-supported Ag, $Au_{25}Ag_{75}$, $Au_{50}Ag_{50}$, $Au_{75}Ag_{25}$, and Au nanoparticles. The XRD is adapted from Ref. [41].

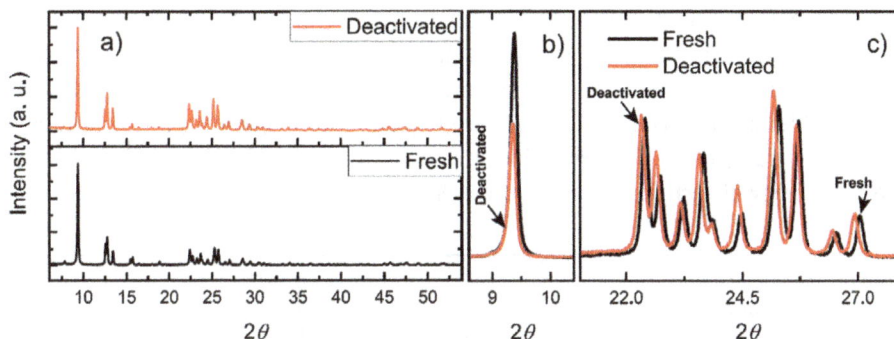

Figure 3.27: (a) X-ray diffraction (XRD) pattern of fresh (black) and fully deactivated (red) for H-Ferrerite zeolite and zoomed key region, **(b)** from 2θ = 8–11, and **(c)** from 2θ = 21–28. The XRD is adapted from Ref. [42].

3.7.3 X-ray photoelectron technique

The XPS is a quantitative analytical technique involving X-rays irradiating a sample to excite electrons from the material's surface (Figure 3.28a). As the size of the matter decrease, a larger percentage of the atoms are formed at the surface. When working with nanomaterials, the properties of the material can be dominated by the surface properties. The results from the XPS analysis gives information about the surface chemistry of the top 10 nm of a sample, based on the photoelectric effect resulting from the emission of electrons from a sample following its irradiation by a beam of X-ray in ultra-high vacuum ($<10^{-8}$ mbar). Two types of peaks could be observed in XPS spectra primary (photoelectron peaks), and secondary (satellites and Auger peaks).

The photoelectron peaks, normally appear as sharp and intense. Dominant and narrow peaks generally appear as shown in the XPS spectrum of a sample containing Cu and Ti oxides (Figure 3.29a). The width of these peaks is determined by the convolution of the width of the X-ray line and the lifetime of the "hole" generated by the photoionization process, which is in the range of the first few atomic layers (1 nm to 10 nm). The photo electrons, represents the observed peaks have not lost any kinetic energy, which means they reach to the detector and contribute to the photoelectron spectra. Another characteristic of photoelectron spectra is that the peaks at higher binding energies (1–4 eV) tend to be less intense and wider than those at lower binding energies. Additionally, the width of the peaks in insulating materials is typically greater than that in conducting materials, which can affect the interpretation of the data.

The basic idea of XPS involves the passing of X-ray irradiation (hv) through the sample, which could lead to expelling of electrons out of the core energy level of the atom, e.g. 1s of O atom. The energy of these ejected electrons provides information about the material's surface chemical composition and electronic state. XPS is widely employed to analyze the properties of solid surfaces and interfaces in different fields such as materials science, chemistry, and physics.

Electrons at various energy levels can be excited depending on the energy of the incident photon. The resulting energy levels as a plot of the photoelectron intensity versus the corresponding Ek of the ejected electrons. All orbital levels, except s ($l = 0$) yield doublet in the spectrum with the two possible states characterized by different building energies ($l > 1$). The peaks in the doublet have specific area ratios based on the degeneracy of each spin state (Figure 3.28b).

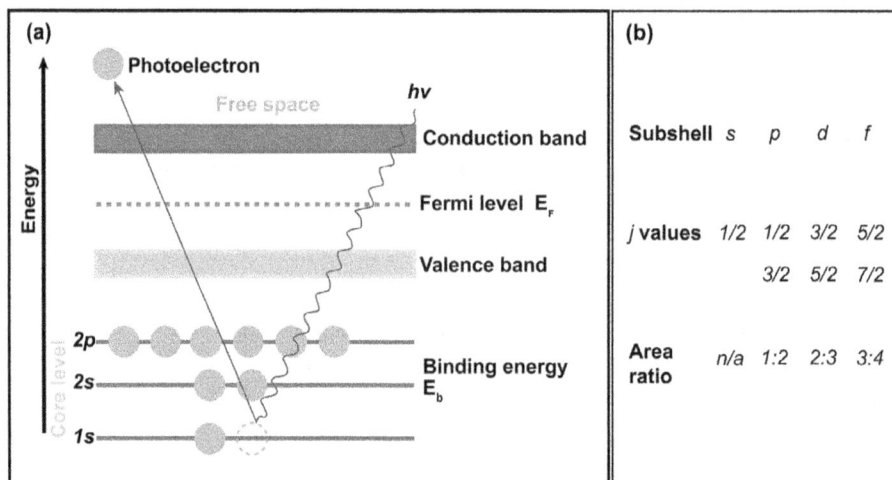

Figure 3.28: (a) Principle of X-ray photoelectron spectroscopy; (b) j Values and area ratio of the peaks in the doublet for the different subshells.

XPS analysis can detect the doublet structure presented in the sample. The doublet structure or multiplet splitting observed in XPS results from the interaction between an electron's spin and angular momentum. When a core electron is excited by X-ray irradiation and emitted (photoemission), its spin 's' can be either parallel or antiparallel to the electron's orbital angular momentum 'l'. For s-levels, the energy values are identical for both spin orientations. At the same time, there is a distinct energy difference between the two states for p, d, or f levels of unfilled shells containing unpaired electrons. The doublet structure depends on the $xnlj$, where 'x' is elements, 'n' is the principal quantum number, 'l' is the orbital angular momentum quantum number (i.e., 0, 1, and 2), j is = $l \pm$. This difference gives rise to two different energy peaks, forming a doublet structure in the XPS spectrum.

In addition to the primary signals in the XPS spectrum, there are also secondary signals known as satellites and Auger peaks. Satellites peaks are caused by a sudden change in Coulombic potential as photoexcited electrons pass through the valence band of the material. Two types of satellite peaks can be observed in high binding energy side peaks: shake-up and shake-off peaks. These secondary signals can be important for quantitation. However, it can be challenging to identify shake-off peaks in

XPS spectra, which can complicate the process of quantification. While the satellite peak of shake-up peaks results from the emission of photoelectrons with reduced kinetic energy. These photoelectrons are emitted from the atom left in an excited energy state after the previous photoelectron ejection. This information plays a role in determining the oxidation state of the atom. On other hand, Auger peaks originate from a process in which an outer-shell electron fills the core-level vacancy created by a previous photoelectron emission. During this process, energy is released, which can cause the ejection of another electron, known as an Auger electron. The Auger peak is named after the scientists who discovered it. The notation used to identify an Auger peak includes information about all the involved levels, such as KL2L3, which refers to the ionization of an electron in the 1s orbital, leaving behind a vacancy filled by an electron from the $2p_{1/2}$ (i.e., L2), resulting in the emission of an electron from the level $2p_{3/2}$ (i.e., L3). A conventional XPS instrument consists of (1) an X-ray tube, where an X-ray beam is generated by bombarding magnesium or aluminum metal target to produce $K\alpha_{1,2}$ radiation, (2) an electron detector consisting of an electron velocity analyzer called a spectrometer, and (3) a pumping system for the high vacuum.

XPS analysis is widely used in the field of heterogeneous catalysis to understand the relationship between catalyst surfaces and their catalytic performance. It allows for the quantitative analysis of a catalyst's elemental composition and identification of the oxidation states of its constituents' elements, including any shifts in the binding energy. Additionally, XPS analysis provides insights into how the physicochemical properties of the catalyst change when exposed to the reaction environment or thermal annealing. By obtaining a comprehensive view of the surface features of the catalyst, researchers can gain a deeper understanding of the initial stages of the catalytic reaction and, therefore, the underlying reaction mechanism.

It is common to first conduct a general survey analysis to get an overview of the elements present in a sample, and then perform a more detailed, narrower scan analysis with higher resolution to accurately examine individual elements in the sample.

Fig. 3.29a, shows the XPS spectrum obtained for CuO/TiO_2 catalyst, which was utilized for CO_2 electrochemical reduction. The XPS survey spectrum confirms the presence of the expected Ti, Cu, and O elements on the surface of the sample. The high-resolution deconvoluted Cu 2p spectrum in Fig. 3.29b reveals the presence of characteristic doublet with Cu $2p_{3/2}$ and Cu $2p_{1/2}$ separated by a distance of 20 eV typical for CuO phase [43]. In addition, appearance of two satellite peaks at higher binding energy further confirm the presence of Cu^{2+} species in CuO. In another example, the O 1s XPS spectrum of Fe-Ti oxide sample shows has one peak with no sign of any doublet, as shown in Figure 3.29c; however, after applying the deconvolution, the broad peak deconvoluted into three different peaks at 528.4 eV, 529.3 eV and 531.1 eV corresponds to the lattice oxygen (O^{2-}) in TiO_2 support, Ti-O-Fe interactive species and hydroxyl groups respectively [44].

The Fig. 3.29d shows for Mo3d XPS spectrum for Sb incorporated 12-ammonium phosphomolybdate (AMPA) catalyst. The major Mo3$d_{5/2}$ and Mo3$d_{3/2}$ doublet appeared

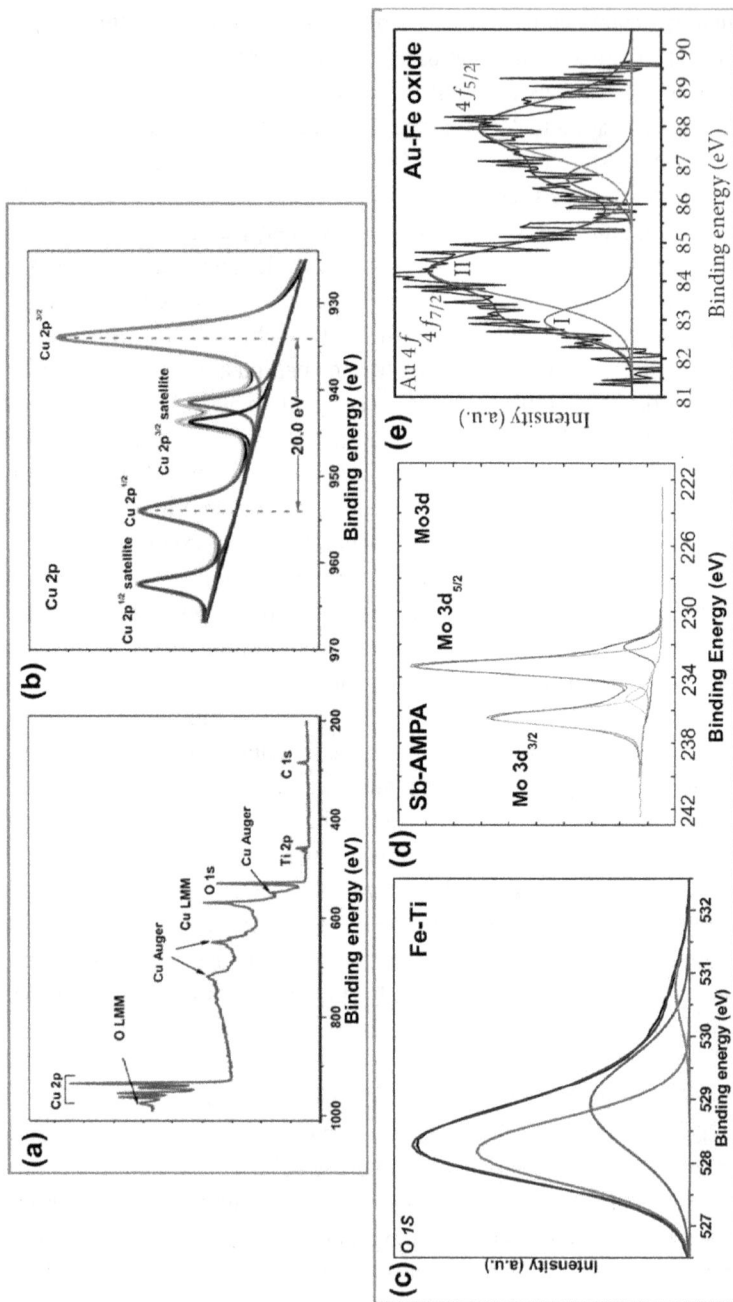

Figure 3.29: (a) XPS survey of a CuO/TiO$_2$ catalyst; **(b)** the high-resolution spectrum of Cu 2p (adapted from Ref. [43]); **(c)** Deconvoluted X-ray photoelectron spectra for Fe-Ti nanoparticles (adapted from Ref. [44]), **(d)** Sb-AMPA sample shows the Mo3d XPS spectrum (adapted from Ref. [45]), and **(e)** Au4f$_{5/2}$ for Au-supported Fe$_2$O$_3$ sample (adapted from Ref. [46]).

at approximately 233.1 eV and 236.1 eV which confirm the presence of Mo^{6+} species in AMPA sample. A minor $Mo3d_{5/2}$ and $Mo3d_{3/2}$ doublet also appeared at approximately 232 eV and 235 eV, which could be assigned to reduced molybdenum (Mo^{5+}) species [45]. The deconvolution of the Au4f energy regions for Au supported Fe_2O_3 sample are shown in the Figure 3.29e. As shown in the Figure 3.29e, the deconvoluted Au4f spectrum shows two doublets corresponding to $Au4f_{7/2}$ and $Au4f_{5/2}$ components. The major doublet for $Au4f_{7/2}$ and $Au4f_{5/2}$ contributions appeared at 84.3 eV and 88.3 eV could be attributed to oxidized Au species, while minor doublet appeared at 83.0 eV and 86.5 eV is due to the presence of metallic Au species on the surface of the sample [46]. These reported XPS spectra for different subshells for different atoms presented on surface of catalyst samples clearly demonstrating the usefulness of the XPS analysis to investigate the different species presented on the catalyst surface, which play a crucial role in the reaction mechanism.

3.8 Electron microscopy

Microscopy involves creating enlarged visual or photographic images of small objects. To achieve this, the microscope must perform three functions: (i) generate a magnified image of the specimen, (ii) separate and resolve the details within the image, and (iii) make these details discernible to the human eye or camera. This technique provides highly localized images and specific information about an extremely small sample, enabling the determination of its structure, morphology, and composition at the atomic scale [15]. Two widely used imaging techniques for visualizing nanoscale features are the scanning electron microscope (SEM) and the transmission electron microscope (TEM). SEM and TEM are both imaging techniques used to visualize samples at the nanoscale. The main difference is how the image is produced: SEM produces images by scanning a focused electron beam across the sample surface, while TEM generates images by passing an electron beam through the sample. SEM can provide higher-resolution surface information, including topography and composition, whereas TEM offers information on internal structures and defects in the sample. It was quickly recognized that operating at higher accelerating voltages offered an advantage, as the wavelength λ of electrons, which have mass m_0, is inversely proportional to the accelerating voltage V:

$$\lambda = \frac{h}{(2m_0eV)^{1/2}} \tag{3.13}$$

where h is Planck's constant, and e is the electron's velocity. The high-quality performance of electron microscopes depends heavily on knowledge of the wavelength of the accelerated electrons used. For example, at an accelerating voltage of 200 kV, the electrons' relativistic wavelength is approximately 0.00251 nm, and at 400 kV, it is about 0.00164 nm. Also, an electron accelerated at a voltage of 1 kV corresponds to a wavelength of approximately 40 pm, whereas an electron accelerated at one megavolt (MV) corresponds to 0.9 pm. Electrons interact intensely with matter (about 100–1,000 times more than X-rays and neutrons), generating various signals in TEM/scanning TEMs. A list of the main signal types generated and utilized in TEM/scanning transmission electron microscope (STEM) is provided in Figure 3.30.

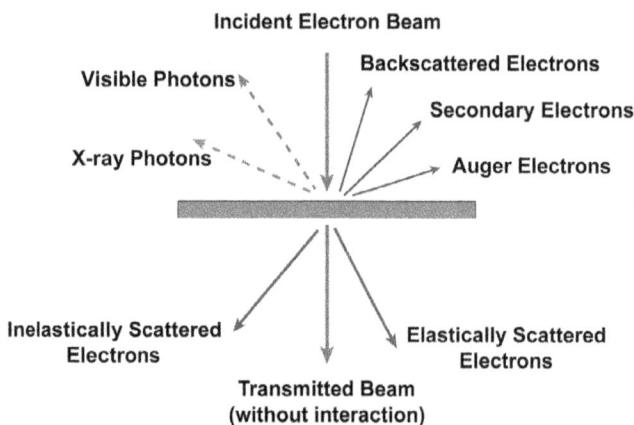

Figure 3.30: The primary types of signals are generated when an electron beam interacts with a specimen in a transmission electron microscope/scanning transmission electron microscope (TEM/STEM) (reproduced from ref. [48]).

3.8.1 Transmission electron microscopy (TEM)

TEM and STEM are complex optical systems that produce high-magnification images of thin specimens. These microscopes use a high-energy electron beam to probe the specimen (100 nm) and collect electrons that have been scattered in the forward direction; that is, they operate in transmission. The magnification in electron microscopes can be tuned between $\times 10^3$ and $\times 10^6$ to resolve distances down to 1.5–1.8 Å. TEM and STEM instruments generally consist of the following: (a) electron gun, (b) probe-forming lenses and apertures, (c) specimen holder, (d) image-forming lenses and apertures, (e) electron detectors, and (f) vacuum system. The main difference between TEM and STEM is in the specimen illumination strategy, parallel beam and convergent beam, and in the detection approach, parallel and serial (or point-by-point), respectively (Figure 3.31).

Images produced by TEM are not perfect due to diffraction at the edges of lenses and various kinds of imperfection in the lenses (aberrations), which causes points within the specimen to appear as diffuse disks surrounded by weak diffraction rings, known as airy disks (as illustrated in Figure 3.32 (resolved airy disks)). The radius of the airy disk, rD, can be defined as the distance between its center and the first minimum. The overlapping of adjacent airy disks limits the point-to-point resolution. Ac-

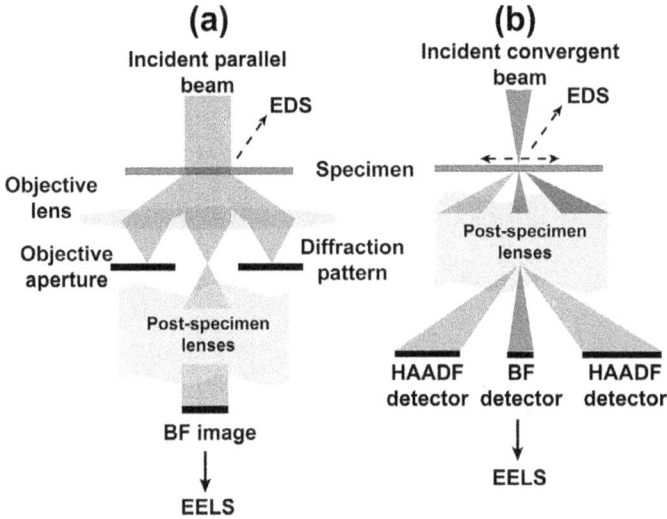

Figure 3.31: Schematic diagram illustrating the illumination and image formation in **(a)** TEM and **(b)** STEM (adapted from ref. [48]).

cording to the Rayleigh criterion, disks can be differentiated when the peaks of their intensity are not closer than the airy disk radius (Figure 3.32; Rayleigh limit, unresolved airy disks). It can be demonstrated that this leads to a resolution limit given by

$$\delta = \frac{0.61\lambda}{\mu \sin \beta} \tag{3.14}$$

where δ is resolution (smallest distance can be resolved), λ is the wavelength (~400 nm for visible light and ~0.004 nm for 100 keV electrons), μ is the refractive index, β is semi-angle of collection, and $\mu \sin \beta$ is also called the numerical aperture. TEMs utilize a beam of electrons to visualize a specimen. The TEM column generates the electron beam, shaped and directed by a series of electromagnetic lenses. The beam passes through the specimen, which may be a thin slice/section of a larger object or a freestanding sample. The beam interacts with the specimen, creating various signals that pass back out of the column to be processed and used to create an image. The resulting image is then displayed on a screen for analysis.

Phase contrast is a fundamental principle in high-resolution TEM (HRTEM) imaging. In basic terms, including an array of reflections centered on the direct, unscattered beam in the objective aperture generates a cross-grating of sinusoidal fringes in the image. These fringes are created from the interference between all the beams incorporated within the aperture. These so-called "lattice fringes" are not necessarily direct images of the crystal structure, as the aberrations of the lenses can alter the relative phases between the beams. However, an image generated by HRTEM can be considered a map of the projected electrostatic potential for electrons along the direc-

Resolved Airy disks   **Rayleigh Limit**   **Unresolved Airy disks**

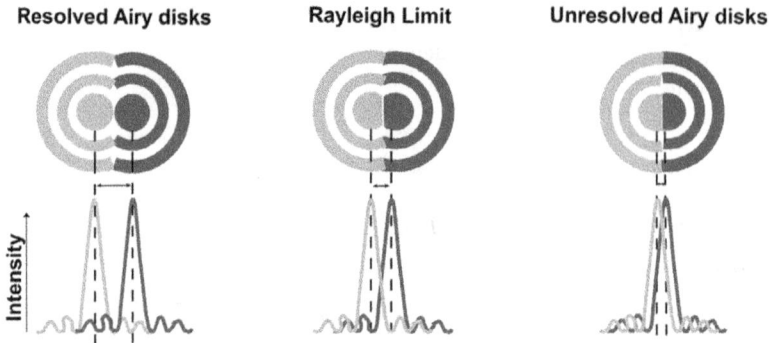

Figure 3.32: Effect of lens aberrations and diffraction on resolution: intensity in airy disks representing images of two object points within a real image.

tion of the incident beam within the specimen. The primary difference between TEM and HRTEM lies in the magnification and resolution achieved. TEM produces images at lower magnifications and provides information on sample morphology, composition, and structure. In contrast, HRTEM offers higher magnification and resolution, allowing one to visualize individual atoms or atomic planes within a sample. This enables HRTEM to provide information on the crystal structure, defects, and interfaces, which are not easily discernible using TEM.

3.8.2 Scanning electron microscopy (SEM)

SEM is a type of microscopy that scans a focused electron beam across a sample less than 1–5 nm to generate high-resolution images of its sample surface morphology, composition, and other characteristics. The essential components of an SEM are (i) the lens system, (ii) the electron gun, (iii) the electron collector, (iv) visual and photographic CRTs, and (v) associated electronic circuitry (Figure 3.33). The electron gun of a SEM gives the microscopist the capability to form a well-defined beam that is characterized by three parameters: beam current (i), beam diameter (d), and beam divergence (α). It works by generating and manipulating a focused electron beam, which is then scanned across the surface of a sample. The electron beam enters the specimen chamber and strikes a single location within the sample. Within the interaction volume between the electron beam and the sample, both elastic and inelastic scattering events occur, giving rise to various detectable signals, such as backscattered electrons, secondary electrons, absorbed electrons, characteristic and continuum X-rays, and cathodoluminescent radiation. By measuring the magnitude of these signals using appropriate detectors, the detector records these signals at each point and then uses them to create an image of the surface properties of the sample, such as local topography, morphology, and surface composition, which can be determined at each individual location and other characteristics with high resolution. The beam needs to be moved from one location to another to cover the desired area, achieved through a scanning system. The scanning motion is performed in both the X and Y directions.

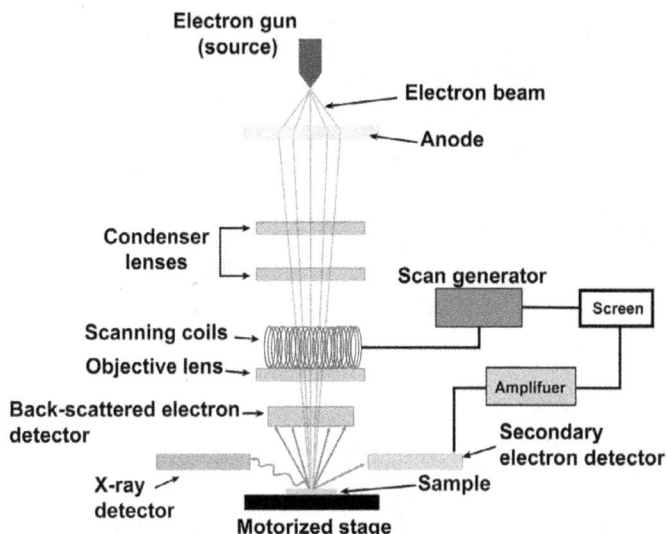

Figure 3.33: Schematic illustrations of the scanning electron microscopes (SEM).

This information can be presented in two primary ways: **(i)** Line scans: the electron beam is scanned across the specimen in the X or Y direction. The same signal the scan generator generates drives a CRT in the same direction. A synchronous ray creates a one-to-one correspondence between points in the "specimen space" and those on the CRT or any other display space. **(ii)** Area (image) scanning: To form the image that we recognize from SEM, the electron beam is systematically scanned over the sample surface in an X-Y grid pattern, while the CRT is similarly scanned in the same X-Y pattern, displaying the electron interaction information. In addition to the elemental analysis system, energy-dispersive spectrometry was employed. This technique can be applied effectively to thick and thin samples for elemental analysis.

3.8.3 SEM and TEM image analysis of catalysts

Figure 3.34a depicts SEM images that unveil the varying surface morphologies of pure ZrO_2 and rGO/ZrO_2 samples. Further, the integration of ZrO_2's morphology is emphasized upon introducing 5 wt.% GO, as seen in the ZG5SRC sample. In HRTEM images, the lattice fringes of ZrO_2 and ZG5SRC are discernible, as shown in Figure 3.34b. The interplanar spacing $d_{(101)}$ of the t-ZrO_2 crystal was distinctly observed in the HRTEM images, yielding a value of 0.30 nm. Clear lattice fringes were visualized, and the distance between adjacent planes was measured to be 0.30 and 0.34 nm, corresponding to (101) planes in t-ZrO_2 and (111) plane in m-ZrO_2, respectively [49].

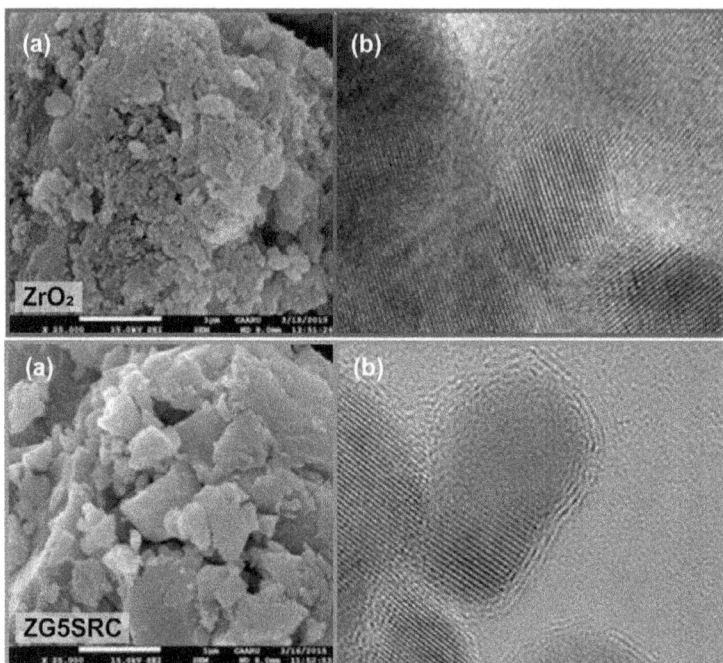

Figure 3.34: **(a)** Scanning electron microscopes (SEM) and **(b)** high-resolution transmission electron microscopy (HRTEM) images of ZrO_2 and ZG5SRC samples (adapted from ref. [49]).

HAADF-STEM, or high-angle annular dark field STEM, is a variant of the STEM technique that images the specimen using high-angle scattered electrons. This technique enhances the contrast of heavy elements and provides information about their elemental composition within the sample. HAADF-STEM is particularly useful in material science and nanotechnology applications. Figure 3.35a depicts the synthesis of ZSM-5 under hydrothermal conditions. Larger zeolites have a diameter of approximately 1 μm, while smaller ones measure around 200 nm. Interestingly, most zeolites display a smooth surface, while a few show a rougher texture. TEM is employed to examine the microstructure of the prepared hollow porous carbon (HPC) catalyst in greater detail. Figure 3.35b reveals the amorphous carbon material obtained through chemical vapor deposition (CVD) preparation conditions. Fundamentally, the carbon material does not adhere to the skeletal structure of the zeolite, nor does it inherit the distinctive elliptical cubic morphology of the zeolite template. This is a result of the slow flow rate of the carbon source gas during the synthesis process, which leads to insufficient carbon deposition and ultimately prevents the formation of HPC via the etching of the zeolite template [50].

Figure 3.35: (a) Scanning electron microscopes (SEM) of synthesized ZSM-5; **(b)** transmission electron microscopy (HRTEM) images of HPC prepared under CVD conditions; and **(c)** HAADF-STEM and **(d)** mapping of various elements in ZSM-5-x (x = 0.05 and 0.2) (adapted from ref. [50]).

Hierarchical zeolites with hollow structures (ZSM-5-x, where x = 0.05 and 0.2) are created by etching ZSM-5 with a NaOH solution. This arises from a selective etching of Si species within the zeolite framework. As the concentration of NaOH solution increases, the contents of O and Si elements within the zeolite decrease. However, the concentration of the Al element is less affected, remaining virtually unchanged after etching (as shown in Figure 3.35c and d).

3.9 Surface area, porosity, adsorption, and diffusion

Nanoporous materials have gained significant usage in adsorption-related applications, finding their way into various fields such as catalysis, pollution control, gas storage and separation, agriculture, pharmaceuticals, and medicine. These materials can be found in household appliances such as water filters, dyeing filters, extractor caps, alcohol tests, and stoppers in medicine tubes. They are even incorporated into certain capsules themselves. The application of adsorption phenomena is not a recent discovery. Egyptians have known and utilized the adsorption properties of porous materials, such as clay and carbon, to purify oil and water since ancient times. A pivotal moment in the wider interest in adsorption research occurred during the First World War when active carbons were employed in gas masks. Freundlich was the first to describe adsorption isotherms mathematically, but theoretical models that are consistent with thermodynamics were developed later by Zsigmondy (1911), Polanyi (1914), and Langmuir (1915). The Brunauer-Emmett-Teller method was introduced in 1938, a significant milestone in adsorption studies. Nowadays, computer-generated models and isotherm reconstruction techniques provide an accurate method for characterizing solid materials. Gas adsorption is frequently employed to determine the surface and porosity of nanoporous materials. Most experiments are conducted using simple gases such as nitrogen, argon, and krypton in a liquid nitrogen cryostat at a temperature of 77 K (-196.15 °C). Other gases (like CO_2) and conditions (273 K for CO_2) have been proposed for identifying porous solids, but they haven't gained widespread use.

Gas adsorption is one of the experimental methods used to measure porous materials' surface area, pore volume, and pore size. The concept of adsorption involves the enrichment of one or more components in an interfacial layer, specifically at the gas/solid interface. The solid is known as the adsorbent, while the gas capable of being adsorbed is called the adsorptive. When a gas is in the adsorbed state, it is called the adsorbate [51]. The term "physical adsorption" or "physisorption" refers to the phenomenon wherein gas molecules adhere to a surface at a pressure lower than its vapor pressure. The attraction between the adsorbed molecules and the surface is relatively weak and definitely not covalent or ionic. In an adsorption experiment, the primary measurement performed is the adsorption isotherm. This refers to measuring the amount of adsorbed versus adsorptive pressure at constant temperature. If the amount required for a monolayer of adsorbate can be determined, then the surface area can be calculated using this information.

3.9.1 Basic concepts in determination of surface area

Various methods can be used to measure adsorption equilibrium. Commonly used techniques include manometric, volumetric, and gravimetric methods.

Gravimetric measurement involves directly measuring the weight change of a solid after adsorption at different gas pressures using specialized adsorption balances (Figure 2.7a). This method is highly accurate and suitable for measuring adsorption at room temperature and above, provided the buoyancy effect is accurately estimated or measured. However, for experiments conducted at 77 K, this gravity-based measurement method has some disadvantages. These include challenges in accurately controlling the sampling temperature and making precise measurements. This is because, while the sample is at a temperature of 77 K, the balance mechanism itself is at room temperature. This results in a significant temperature gradient between the two. Additionally, the sample is held inside a tube and does not directly contact the cryogenic bath, leading to temperature changes between the two, especially at the initial adsorption point. This is further exacerbated by the fact that the sample and bath are in distinct vacuum conditions at this point, causing further temperature variations.

While the **volumetric** or **manometric method** may be less accurate than the gravimetric method, the experimental setup is simpler and cheaper, allowing for direct contact between the sample and the cryostat and enabling better control of the sample temperature (see Figure 2.7b). The advantage of performing adsorption at lower temperatures means that the additional accuracy obtained through gravimetric measurements is typically not significant in 77 K experiments. Despite this, the term "volumetric" is still widely used to describe modern experiments, even though there is no actual measurement of volume change. This misuse is related to early experiments using mercury to measure gas phase volume differences. Modern equipment, on the other hand, utilizes sensitive pressure meters.

3.9.2 Surface area role in catalysis

The general experimental procedure for measuring surface area typically involves four steps: (1) activation, (2) dead space volume calibration, (3) adsorption, and (4) desorption. The optimal amount of sample used for adsorption experiments can vary depending on the measurement method. For the manometric method, around 40 m^2 of surface area is ideal, which equates to approximately 80 mg of sample in a cell for a material with a surface area of about 500 m^2/g. However, for samples with low surface areas, this can lead to a large amount of powder required in the cell. However, for low-surface area samples, this can result in a high amount of powder in the cell, leading to difficulties with diffusion in the sample bed. For high-surface area samples, it's important to ensure minimal amounts in order to prevent any sampling issues. Before experimentation, the sample must be defined and reproducible after being activated or outgassed. The aim is usually to eliminate any physisorbed species (contaminants or humidity, etc.) on the surface without damaging its structure. Therefore, the outgassing temperature is usually chosen within the range where the desorption curve is horizontal. For example, Figure 3.36a shows the quasi-horizontal region of a NaX zeolite. The temperature should be chosen at the beginning of the plateau between 623 K (400 °C) and 1,073 K (800 °C). Additionally, it's important to note that the sample is often outgassed under vacuum conditions, meaning the thermogravimetric analysis (TGA) curve is obtained under a dry atmosphere.

The outgassing of the sample often occurs under a vacuum, and the vacuum should be applied slowly to prevent the powdered sample from entering the vacuum pump. Once under vacuum, a suitable outgassing temperature can be set. Again, TGA curves can be used to determine if intermediate temperatures can be utilized. This can occur when there is a significant amount of weakly adsorbed species. Leaving the

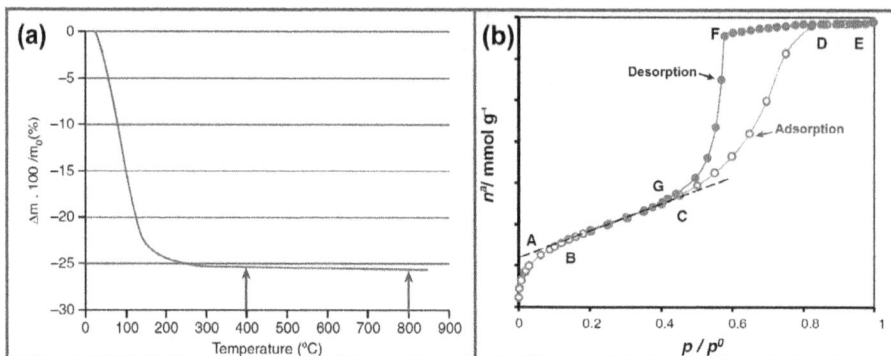

Figure 3.36: (a) The thermogravimetric analysis for 13× zeolite sample and **(b)** the adsorption-desorption isotherm obtained with N_2 on a porous glass at 77 K (adapted from ref. [52]).

sample at approximately 353 K (80 °C) for some time can help prevent sample degradation due to the rapid loss of these weakly binding species. After outgassing, the sample is weighed again and placed in the adsorption manifold. The first step of determining the dead space is performed using helium. The dead space includes part of the sample cell and the manifold. Since different sample cells are not the same and sample volume varies in each experiment, the dead space should be determined before each experiment. Helium is used because it is believed not to be adsorbed. The actual measurement of dead space using helium is performed similarly to isothermal adsorption experiments. The adsorption measurement involves introducing a known quantity of the adsorption gas into the reference volume V_{ref}. The reference volume is then connected to the sample volume V_{sample} until a balance is achieved between the sample and the gas. The attainment of equilibrium can be evaluated in several ways. One method involves tracking the pressure above the sample and considering it balanced if the pressure change remains below a specified limit within a certain timeframe. Alternatively, a simple time limit can be set. This equilibrium is an important aspect of the experiment, as results can be misinterpreted if a poor equilibrium is obtained at individual measurement points. In the case of nitrogen adsorption at normal liquefaction temperature (77.4 K), the adsorption isotherm (Figure 3.36(b)) represents the equilibrium state of the adsorbent from zero to the saturation vapor pressure of the adsorptive. However, each isotherm is unique to the adsorbate-adsorbent pair being studied and should take into consideration factors such as gas polarity and the solid surface chemistry, including hydrogen content, presence of catalysts, and free metal sites. However, the primary factor to consider is the size and volume of the pore. Two types of pores, micropores (with pore openings between 0.4 and 2 nm) and mesopores (with pore openings between 2 and 50 nm) are commonly found in solids. Porous materials are classified into three categories based on the size of their pores (r_{pore}): microporous for $r_{pore} < 2$ nm, mesoporous for 2 nm $< r_{pore} < 50$ nm, and macroporous for $r_{pore} > 50$ nm. Microporous crystals (also known as zeolites) with over 130 distinct framework-type structures have been documented [51]. In both instances, materials with nanoporous structures with nanometer dimensions are commonly classified as nanoporous. According to the IUPAC recommendation, the isotherm is usually presented graphically as the amount of adsorption per gram of adsorbent as a function of the equilibrium pressure. When performed at a temperature lower than the triple point, the amount of adsorbed is represented as a function of the relative equilibrium pressure (P/P_0), where P_0 denotes the saturation vapor pressure of the adsorptive.

For instance, the isotherm in Figure 3.36b illustrates adsorption-desorption branches. Different points on the isotherm provide initial insights into the adsorption process. **Point A** marks the back-extrapolation to $P/P_0 = 0$ of the quasi-linear region B-C. The adsorption amount at this point A can, in certain cases, be associated with the volume of micropores. **Point B** signifies the start of a quasi-linear region of the isotherm, extending to **point C**. Point B is the point where a gaseous monolayer forms on the solid's surface. Initially, the amount of adsorption at this point can be estimated to determine an approximate surface area. The quasi-linear region between B and C corresponds to the formation of a multilayer. The region between C and D on the isotherm reflects an upturn due to capillary condensation in mesoporous solids as the pores get filled. The region between D and E on the isotherm corresponds to the final stage of pore filling, and the adsorption amount can be associated with the total pore volume. **Point F** on the desorption branch of the isotherm marks the commencement of pore emptying, and the relative pressure where this occurs can be linked to the pore size. **Point G** denotes the juncture where the desorption isotherm reconnects with the adsorption branch. A certain gas at a specific temperature will have a characteristic minimum value for this point, known as the point of meniscus instability. For a given gas and temperature, there exists a minimal pore size below which the liquid meniscus spontaneously breaks, permitting full desorption from any mesopores. For nitrogen at 77 K, this occurs at $P/P_0 = 0.42$, whereas for argon at the same temperature, meniscus instability happens at $P/P_0 = 0.28$. The details about different adsorption isotherms and hysteresis are presented in Chapter 2. For example, in Figure 3.37a the textural properties of synthesized NiSil samples, both bare and ZnO deposited, were investigated using N_2 physisorption analysis. The obtained isotherms demonstrated type V characteristics, consistent with mesoporous materials as classified by IUPAC. The observed H_2-type hysteresis loops indicate that the samples possess disordered bottleneck-shaped pores. The pore size distribution analysis (Figure 3.37b) indicated a relatively minor peak at 14 Å and a major peak around 20 Å. The presence of large, open NiSil nanotubes in the catalyst resulted in an average pore size of approximately 19 Å [53].

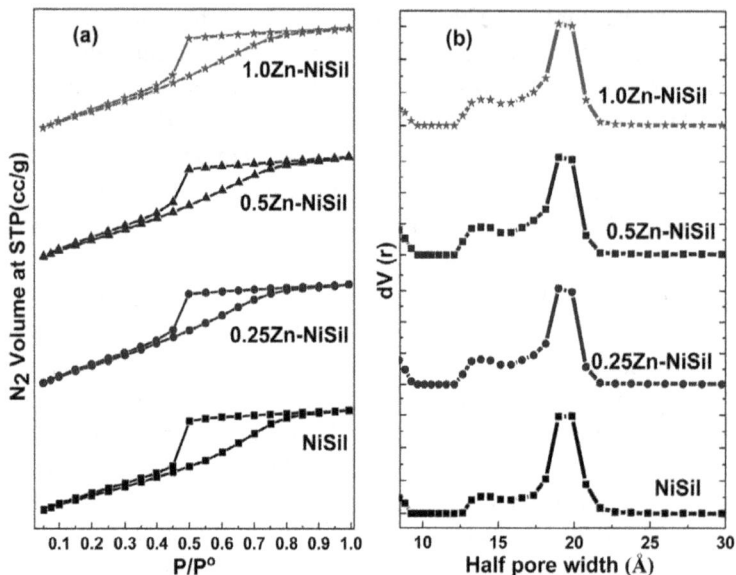

Figure 3.37: (a) N$_2$ adsorption-desorption isotherms and **(b)** pore size distribution patterns of the samples (adapted from ref. [53]).

3.10 Thermal techniques

Thermal analysis can measure the physical and chemical properties of materials as a function of temperature. However, in practice, the term "thermal analysis" often refers to specific properties such as enthalpy, heat capacity, mass, or coefficient of thermal expansion. For instance, measuring the change in weight of oxy salts or hydrates as they decompose on heating is an example of thermal analysis. The two primary thermal analysis techniques are Thermogravimetric analysis (TGA), which automatically records a sample's weight change as a function of either temperature or time. The other principal technique is differential thermal analysis (DTG), which measures the difference in temperature, ΔT, between a sample and an inert reference material as a function of temperature. DTG, therefore, identifies changes in heat content. Since catalysts play a significant role in various reactions due to their acid-base sites, it is essential to accurately determine the number, strength, and strength distribution of these sites. Thermal techniques, such as temperature-programmed desorption (TPD), temperature-programmed reduction (TPR), and temperature-programmed oxidation (TPO), can be used to determine these parameters.

3.10.1 Differential thermal analysis (DTA)

DTA, which stands for "differential thermal analysis," is a technique used to measure the difference in temperature between two substances while the temperature is being programmed. This is done by placing one material, the sample, and another material, the reference, in a heating vessel in crucibles, controlling the atmosphere in the heating vessel. DTA analysis helps to detect and quantify the energy changes associated with physical and chemical changes that happen in the sample material due to heating (Figure 3.38a).

3.10.2 Differential scanning calorimetry (DSC)

The principle behind differential scanning calorimetry (DSC) is similar to DTA because it is a temperature-programmed technique. Still, instead of measuring the difference in temperature between the sample and a reference, it measures the difference in heat flow (or thermal power) between them. This is achieved using a differential heat flux sensor and controlling the atmosphere in the heating vessel (Figure 3.38b).

3.10.3 Thermogravimetry analysis (TGA)

TGA is an analytical technique used to measure the change in mass of a sample as a function of temperature or time while it is subjected to a programmed temperature profile or a controlled atmosphere. TGA is useful for analyzing processes involving mass changes including drying, desorption, reduction, degradation in active atmospheres, and other processes related to mass loss or gain. TGA is particularly valuable for studying solid-gas systems such as adsorption, oxidation, and other reactions involving gas-solid interactions. TG along with DTA or DSC offers several key benefits, including

1. Identification of thermal events: TGDTA or TGDSC can detect and quantify of exothermic and endothermic processes such as phase transitions, dehydration, oxidation, reduction, decomposition, and volatilization.
2. Determination of material composition and purity: TGDTA or TGDSC can be used to determine the percentage of volatile and non-volatile components in a sample, providing insights into the material's composition and purity.
3. Evaluation of thermal stability: TGDTA or TGDSC can assess a material's thermal stability and resistance to thermal degradation by monitoring the change in mass over a specific temperature range.

For example, the thermogravimetric profiles (under nitrogen) of Nanoporous phosphotungstate organic-inorganic hybrid materials synthesized from sodium tungstate and mono-n-dodecyl phosphate (MDP). The thermal decomposition behavior of both the MDP

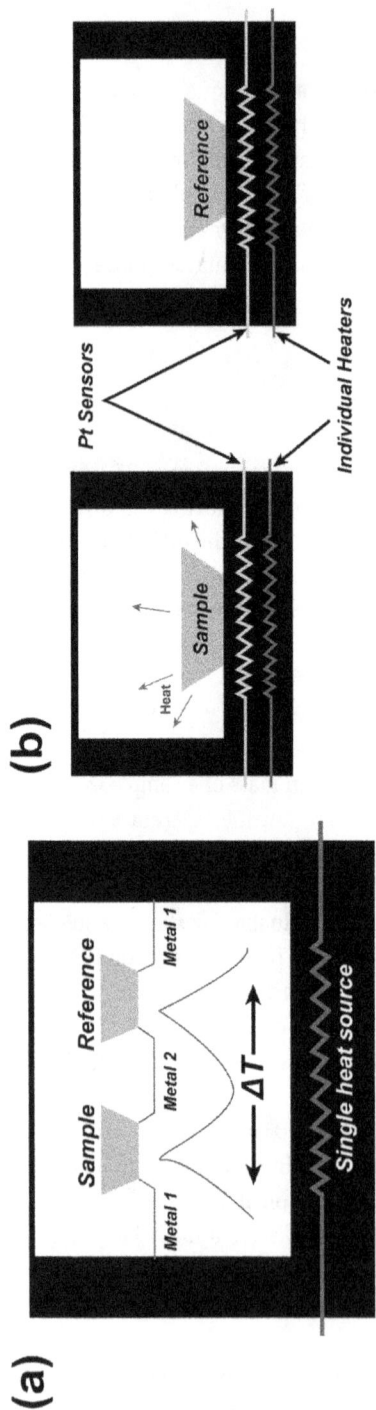

Figure 3.38: (a) Differential thermal analysis (DTA) and **(b)** differential scanning calorimetry (DSC).

precursor and the MDP-polyoxometallite hybrid material were studied through TGA (Figure 3.39). The MDP sample experienced exothermic weight losses at 120 and 230 °C, attributed to the release of water and decomposition of the MDP hydrocarbon tail and total loss was 70 wt.% by 500 °C. The MDP-POM sample exhibited a weight loss at 285 °C, resulting in a total weight loss of 30 wt.% at a temperature of 500 °C. The thermal stability of the MDP-POM material appears to be enhanced due to direct interaction between MDP and tungstate clusters [54].

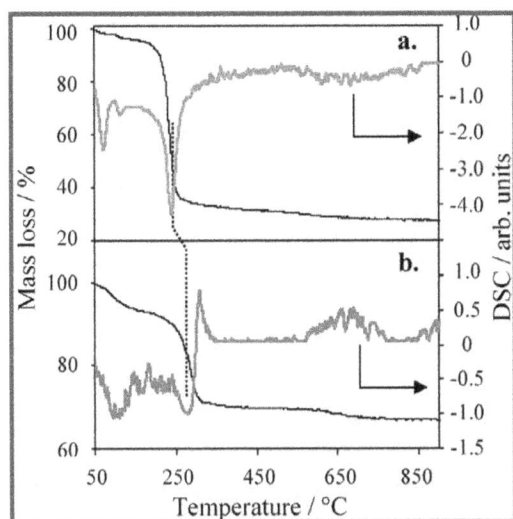

Figure 3.39: TGA and DSC measurements for **(a)** MDP and **(b)** MDP-POM hybrid sample (adapted from ref. [55]).

3.10.4 Temperature-programed techniques

The term "temperature-programmed techniques" TPX is widely used in the field of catalysis, where X represents the process studied, such as desorption (TPD), reduction (TPR), oxidation (TPO), and sulfidation (TPS). TPX techniques are employed to study different phenomena occurring on the catalysts surfaces as a function of temperature. The apparatus of the TP-techniques includes in general of **(a)** a temperature controller, **(b)** a gas delivery system, **(c)** a furnace, and **(d)** a sample holder adapted to one or more of the gravimetric, calorimetric, or spectroscopic (see Figure 3.40).

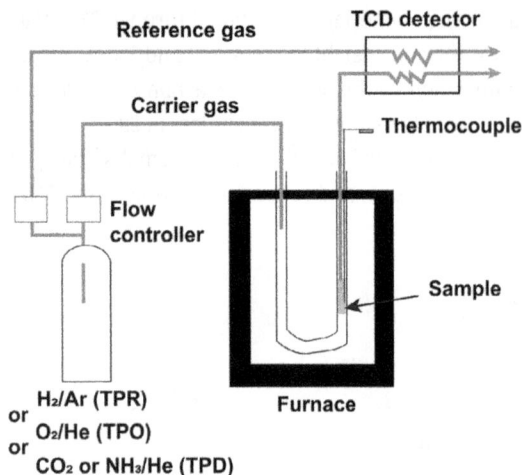

Figure 3.40: The temperature-programmed setup.

3.10.5 Temperature-programmed desorption (TPD)

TPD commonly used to study the nature and amount of species that are desorbed from an initially impregnated substrate, as a temperature program is run. TPD is predominantly employed to characterize the acidic or basic properties of surfaces and decompose adsorbed species at specific temperatures. The concept of TPD consists of three phases: (i) pretreatment, (ii) adsorption, and (iii) desorption.

- **Pretreatment:** Prior to the analysis, the sample is brought to a low pressure or inert gas flow and a temperature that is sufficient to "clean" the surface without causing any damage or degrading the sample's structure.
- **Adsorption:** In the subsequent step, one or several species are adsorbed onto the sample surface at low temperature (commonly room temperature, although occasionally subambient) until achieving a specific surface coverage or reaching saturation.
- **Desorption:** The sample is typically heated in a controlled manner (often linearly) while gas (either inert or otherwise) carries the desorbed molecules into a detector, or under ultra-high vacuum conditions to prevent the molecules from being reabsorbed. The desorption process is sometimes performed in two steps to distinguish between physisorbed and chemisorbed species. Additionally, it is usual to observe the products of decomposition of the initially adsorbed molecules at high temperatures.

TPD is widely used for the characterization of acidic/basic sites in gas-solid interactions, and it provides information on the amount and energy of these sites. Various gas probes are used, but predominantly four probes are commonly employed: NH_3, and pyridine for testing acid sites, while SO_2, and CO_2 for testing basic sites. Other probes are used for specific measurements such as CO for testing strong metal-support interaction. An example of investigating the acidic property of ZSM-5 and zeolite-like materials (SAPO-34) using NH_3-TPD is shown in Figure 3.41a. The TPD-NH_3 profile indicates the presence of two desorption peaks corresponding to weak and strong acid sites in both materials (ZSM-5 and SAPO-34). Notably, the profile reveals that SAPO-34 possesses stronger acid properties when compared to ZSM-5 [56]. The CO_2 desorption profile for the layered double hydroxide (CuMgAl) sample is presented in Figure 3.41b. The CuMgAl sample shows three desorption peaks corresponding to weak, moderate, and strong basic sites. It is broadly acknowledged that weakly basic sites are associated with OH structural groups (Brønsted basic sites), moderate basic sites are attributed to magnesium-oxygen pairs (Lewis basic sites), and strong basic sites are linked to low-coordination oxygen atoms. The total number of basic sites are attributed to the presence of −OH and Mg−O groups [57].

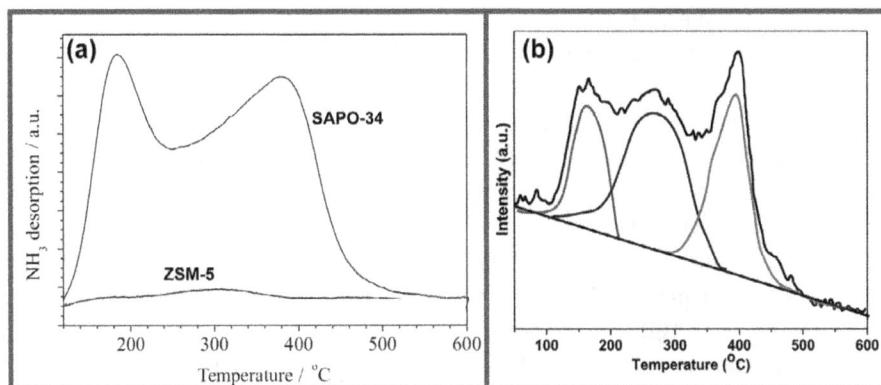

Figure 3.41: (a) NH_3-TPD profile to investigate the acid sites for the SAPO-34 and ZSM-5 samples (adapted from ref. [56]); (b) CO_2-TPD profile to study the basicity sites for the layered double hydroxide (CuMgAl) (adapted from ref. [57]).

3.10.6 Temperature-programmed reduction and oxidation (TPR and TPO)

The thermal analysis techniques which are frequently used for characterizing catalysis are temperature-programmed desorption (TPD), temperature-programmed reduction (TPR) and temperature-programmed oxidation (TPO). Temperature-programmed techniques are simple to accomplish and cost-effective. TPD, TPR and TPO are similar to one another, using same equipment with slight modifications. The TPR and TPO techniques enables us to investigate the redox properties of a catalysts. However, it must be noted that these methods involve exposing the samples with reacting gases at high temperatures and often do not just characterize the surface but also the bulk of the material as well (at least partially in some cases).

The H_2-TPR analysis involves utilization of a reducing gas such as H_2 mixture [diluted in an inert gas (He or Ar), usually 10%H_2 in Ar] flows through a metal oxide sample. The initial temperature is normally at room temperature (25 °C). Then the temperature of the sample is increased at a constant rate (normally 10 °C/min) and, as the reduction begins, and H_2 is consumed from the gas mixture, which is detected by a thermal conductivity detector (TCD). When reduction completes, no more H_2 gas is consumed and the thermal conductivity of the gas mixture from the sample does not change. Several reduction peaks could be observed in the reduction pattern because reduction likely will be initiated at various thermal energy levels.

The TPO technique normally involves utilization of diluted O_2 gas in inert gas and used to determine the quantity of reduced species, however the difficulty with TPO is achieving undesired total oxidation in some cases. More frequently, TPO is used in applications such as investigating the kinetics of coking process, burning off the carbon on catalyst surface and determination of the different forms of carbonaceous materials present in the catalysts.

TPD is a dynamic technique to investigate the nature of sites available for chemisorption of different species. This method involves pre-treating the sample with inert gas in the range of 120–250 °C to remove any adsorbed species from surface of the sample. Then, a selected gas (H_2, NH_3 and CO_2) or vapor (pyridine) is pass through the sample until the surface saturation with chemisorbed species is achieved. Then, the sample flushed with an inert gas to remove any physisorbed species. After that the temperature of the sample is increased at a controlled rate (normally 10 °C/min) while a constant flow of inert gas is pass through the sample. The desorbed molecules in inert gas are monitored by using TCD. The TCD signal is proportional to the quantity of molecules desorbed as temperature controlled by using thermocouple and temperature controller. The desorbed amount at specific temperatures provide information about the strength, number and heterogeneity of the active sites for chemisorption. The TCD signal is usually plotted as quantity desorbed versus temperature.

H_2-TPR, or hydrogen TPR, can provide information about the reproducibility of catalysts. By examining the H_2-TPR patterns of the samples, as shown in Figure 3.42a, it is evident that pure ZrO_2 support did not exhibit any reduction peaks within the studied temperature range of 25–600 °C. However, two distinct reduction peaks at low and high temperatures (peaks 1 and 2) were observed for the 10-FeZr and 20-FeZr samples. Indeed, the observation of two distinct reduction peaks suggests that at least two different types of iron oxide species are present on the surface of the ZrO_2 catalyst. Peak 1 of the low-temperature reduction peaks is associated with iron oxide that has a weak interaction with ZrO_2 support, suggesting that these species can be easily reduced at lower temperatures. On the other hand, Peak 2, occurring at higher reduction temperatures, indicates the presence of iron oxide species with a strong interaction with the ZrO_2 support. This strong bond between the iron oxide and ZrO_2 contributes to the high reduction temperatures required to break these bonds. This finding indicates a complex interplay between iron and zirconia, which may influence the catalytic behavior of the material [58].

In case of the O_2-TPO, the primary benefit of TPO is its ability to provide quantitative information about the oxidation of materials under well-defined and controlled conditions. This allows for the determination of oxidation mechanisms, kinetic parameters, and stability of materials under different oxygen partial pressures and temperatures. TPO can also evaluate the effectiveness of oxidation inhibitors and antioxidant additives. Let's consider an example. Figure 3.42b shows the profiles obtained during the TPO analysis of the pre-reduced CuCe-3 sample. It was found that no O_2 was consumed at $T = 0$ °C, which indicated no consumption of O_2 during first 5 min when a catalyst sample was exposed to an O_2 stream. However, it was found that most of the examined catalyst samples could be reoxidized up to $T = 400$ °C. The reoxidation process of the CuCe-3 sample belonged to $Cu^+ \rightarrow Cu^{2+}$ oxidation steps, which are lower than the corresponding peak areas belonging to $Cu^0 \rightarrow Cu^+$ oxidation steps [54].

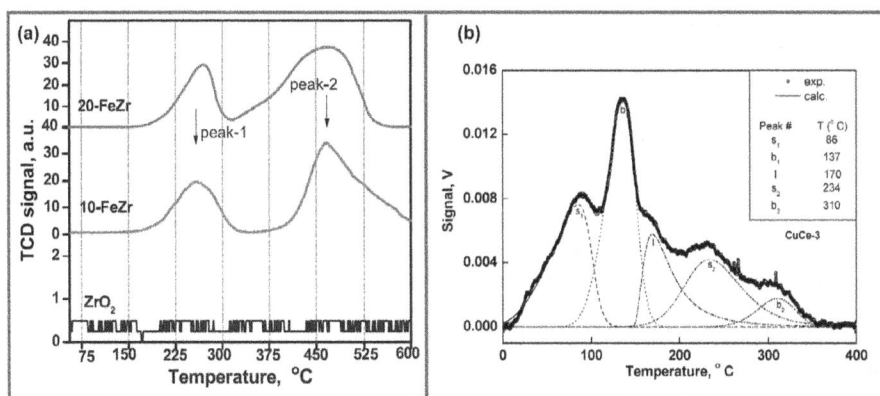

Figure 3.42: (a) H_2-TPR profile to reduction of studied samples (adapted from ref. [58]); **(b)** O_2-TPO profile to study the oxidation of the CuCe sample (adapted from ref. [54]).

3.11 References

[1] M. Che and J. Vedrine, General introduction: in *Characterization of Solid Materials and Heterogenous Catalysts*, eds. M. Che and J. Vedrine, WILEY-VCH, France, 2012, vol. 1, pp. XXXI–XLIII.

[2] D. Barthomeuf, Spectroscopic investigation of zeolite properties: in *Zeolite Microporous Solids: Synthesis, Structure, and Reactivity*, eds. E. G. Derouane, F. Lemos, C. Naccache and F. Ribeiro, Kluwer Academic Publishers, Estoril, 1992, pp. 193–223.

[3] H. G. Karge, Chapter 15- Characterization by IR spectroscopy: in *Verified Syntheses of Zeolitic Materials*, ed. H. Robson, Elsevier Science B, Netherland, 1st edn., 2001, pp. 69–71.

[4]   C. Li and Z. Wu, Microporous materials characterized by vibrational spectroscopies: in *Handbook of Zeolite Since and Technology*, eds. S. M. Auerbach, K. Carrado and P. Dutta, Marcel Dekker Inc, New York, NY, 1st edn., 2003, pp. 417–507.

[5]   S. K. Pirutin, S. Jia, A. I. Yusipovich, M. A. Shank, E. Y. Parshina and A. B. Rubin, Vibrational spectroscopy as a tool for bioanalytical and biomonitoring studies, *Int. J. Mol. Sci.*, 2023, **24**, 6947.

[6]   M. Manzoli. Boosting the characterization of heterogeneous catalysts for H_2O_2 direct synthesis by infrared spectroscopy, *Catalysts*, 2019, **9**, 30.

[7]   F. Rouessac and A. Rouessac. *Chemical analysis: Modern instrumenation methods and techniques*, Wiley & Sons, 2007.

[8]   O. Lahodny-Sarc and J. L. White. Infrared study of aluminum-deficient zeolites in the region 1300 to 200 cm^{-1}, *J. Phys. Chem.*, 1971, **75**, 2408–2410.

[9]   A. Mekki, A. Benmaati, A. Mokhtar, M. Hachemaoui, F. Zaoui, H. Habib Zahmani, M. Sassi, S. Hacini and B. Boukoussa. Michael addition of 1,3-dicarbonyl derivatives in the presence of zeolite Y as an heterogeneous catalyst, *J. Inorg. Organomet. Polym. Mater.*, 2020, **30**, 2323–2334.

[10] M. Khamidun, M. Ali Fulazzaky, A. Al-Gheethi, U. Md Ali, K. Muda, T. Hadibarata and M. Mohammad Razi. Adsorption of ammonium from wastewater treatment plant effluents onto the zeolite; A plug-flow column, optimisation, dynamic and isotherms studies, *Int. J. Environ. Anal. Chem.*, 2022, **102**, 8445–8466.

[11] A. H. Jawhari, N. Hasan, I. A. Radini, K. Narasimharao and M. A. Malik. Noble metals deposited $LaMnO_3$ nanocomposites for photocatalytic H_2 production, *Nanomaterials*, 2022, **12**, 2985.

[12] F. Thibault-Starzyk and F. Maugé, Infrared spectroscopy: in *Characterization of Solid Materials and Heterogeneous Catalysts*, Weinheim, 2012, pp. 1–48.

[13] C. Binet, A. Jadi, J. Lamotte and J. C. Lavalley, Use of pyrrole as an IR spectroscopic molecular probe in a surface basicity study of metal oxides, *J. Chem. Soc. Faraday Trans.*, 1996, **92**, 123–129.

[14] Y.-L. Tsai, E. Huang, Y.-H. Li, H.-T. Hung, J.-H. Jiang, T.-C. Liu, J.-N. Fang and H.-F. Chen. Raman spectroscopic characteristics of zeolite group minerals, *Minerals*, 2021, **11**, 167.

[15] G. Wolfgang, K. Wolfgang and M. Martin. *Catalysis at Surfaces*, Walter de Gruyter GmbH, 2023.

[16] S. Ye, J. Sun, X. Yi, Y. Wang and Q. Zhang. Interaction between the exchanged Mn^{2+} and Yb^{3+} ions confined in zeolite-Y and their luminescence behaviours, *Sci. Rep.*, 2017, **7**, 46219.

[17] J. Chen, G. Peng, T. Liang, W. Zhang, W. Zheng, H. Zhao, L. Guo and X. Wu. Catalytic performances of Cu/MCM-22 zeolites with different Cu Loadings in NH_3-SCR, *Nanomaterials*, 2020, **10**, 2170.

[18] M. Sun, Y. Li, B. Zhang, C. Argyropoulos, P. Sutter and E. Sutter, Plasmonic Effects on the Growth of Ag Nanocrystals in Solution, *Langmuir*, 2020, **36**, 2044–2051.

[19] S. N. Mailu, T. T. Waryo, P. M. Ndangili, F. R. Ngece, A. A. Baleg, P. G. Baker and E. I. Iwuoha, Determination of anthracene on Ag-Au alloy nanoparticles/overoxidized-polypyrrole composite modified glassy carbon electrodes, *Sensors*, 2010, **10**, 9449–9465.

[20] E. Borodina, H. Sharbini Harun Kamaluddin, F. Meirer, M. Mokhtar, A. M. Asiri, S. A. Al-Thabaiti, S. N. Basahel, J. Ruiz-Martinez and B. M. Weckhuysen. Influence of the reaction temperature on the nature of the active and deactivating species during methanol-to-olefins conversion over H-SAPO-34, *ACS Catal.*, 2017, **7**, 5268–5281.

[21] H. S. Kamaluddin, S. N. Basahel, K. Narasimharao and M. Mokhtar. H-ZSM-5 materials embedded in an amorphous silica matrix: Highly selective catalysts for propylene in methanol-to-olefin process, *Catalysts*, 2019, **9**, 364–382.

[22] H. Sharbini Kamaluddin, X. Gong, P. Ma, K. Narasimharao, A. Dutta Chowdhury and M. Mokhtar. Influence of zeolite ZSM-5 synthesis protocols and physicochemical properties in the methanol-to-olefin process, *Mater. Today Chem.*, 2022, **26**, 101061.

[23] E. C. Nordvang, E. Borodina, J. Ruiz-Martínez, R. Fehrmann and B. M. Weckhuysen. Effects of coke deposits on the catalytic performance of large zeolite H-ZSM-5 crystals during alcohol-to-

hydrocarbon reactions as investigated by a combination of optical spectroscopy and microscopy, *Chem. – A Eur. J.*, 2015, **21**, 17324–17335.

[24] E. Borodina, F. Meirer, I. Lezcano-Gonzalez, M. Mokhtar, A. M. Asiri, S. A. Al-Thabaiti, S. N. Basahel, J. Ruiz-Martinez and B. M. Weckhuysen. Influence of the reaction temperature on the nature of the active and deactivating species during methanol-to-olefins conversion over H-SSZ-13, *ACS Catal.*, 2015, **5**, 992–1003.

[25] H. A. Mahmoud, K. Narasimharao, T. T. Ali and K. M. S. Khalil. Acidic peptizing agent effect on anatase-rutile ratio and photocatalytic performance of TiO_2 nanoparticles, *Nanoscale Res. Lett.*, 2018, **13**, 48.

[26] T. ALI, K. Narasimharao, S. Basahel, M. Mokhtar, E. Alsharaeh and H. Mahmoud. Template assisted microwave synthesis of rGO-ZrO_2 composites: Efficient photocatalysts under visible light, *J. Nanosci. Nanotechnol.*, 2019, **19**, 5177–5188.

[27] K. Narasimharao and H. S. Kamaluddin. Adsorption of methylene blue and metachromasy over analcime zeolites synthesized by using different Al precursors, *Mater. Today Chem.*, 2023, **32**, 101675.

[28] A. G. Stepanov, Chapter 4- Basics of solid-state NMR for application in zeolite science: Material and reaction characterization: in *Zeolites and Zeolite-Like Materials*, eds. B. F. Sels and L. M. B. T.-Z. and Z.-L. M. Kustov Elsevier, Amsterdam, 2016. pp. 137–188.

[29] L. Gladden, M. Lutecki and J. McGregor, Nuclear magnetic resonance spectroscopy: in *Characterization of Solid Materials and Heterogeneous Catalysis*, eds. J. Thomas and W. Thomas, WILEY-VCH, 2nd edn., Weinheim, 2015, pp. 289–342.

[30] K. Mukhopadhyay, A. B. Mandale and R. V. Chaudhari. Encapsulated HRh(CO)(PPh3)$_3$ in microporous and mesoporous supports: Novel heterogeneous catalysts for hydroformylation, *Chem. Mater.*, 2003, **15**, 1766–1777.

[31] K. Narasimharao. Effect of Ti incorporation on structure and oxidation activity of ammonium salt of 12-molybdophosphoric acid catalysts, *Int. J. Electrochem. Sci.*, 2013, **8**, 9107–9124.

[32] K. Narasimharao, B. H. Babu, N. Lingaiah, P. S. S. A. I. Prasad and S. A. Al-Thabaiti. Ammoxidation of 2-methyl pyrazine on supported ammonium salt of 12-molybdophosphoric acid catalysts: The influence of nature of support, *J. Chem. Sci.*, 2014, **126**, 487–498.

[33] K. Narasimharao, D. R. Brown, A. F. Lee, A. D. Newman, P. F. Siril, S. J. Tavener and K. Wilson. Structure–activity relations in Cs-doped heteropolyacid catalysts for biodiesel production, *J. Catal.*, 2007, **248**, 226–234.

[34] Introduction to X-ray Diffraction, https://www.princetoninstruments.com/learn/x-ray-scattering/introduction-to-x-ray-diffraction#:~:text=X—ray diffraction is a,details on the atomic level., (accessed 28 August 2023).

[35] S. Taha. *Birkbeck*, College London, 2010.

[36] C. Michel and C. V. Jacques. *Characterization of solid materials and heterogeneous catalysis: From structure to surface reactivity*, Wiley-VCH Verlag GmbH & Co. KGaA, 1st edn., 2012.

[37] M. Singh, R. Kalaivani, S. Manikandan, N. Sangeetha and A. K. Kumaraguru. Facile green synthesis of variable metallic gold nanoparticle using Padina gymnospora, a brown marine macroalga, *Appl. Nanosci.*, 2013, **3**, 145–151.

[38] K. Nagaraj, P. Thangamuniyandi, S. Kamalesu, M. Dixitkumar, A. K. Saini, S. K. Sharma, J. Naman, J. Priyanshi, C. Uthra, S. Lokhandwala, N. M. Parekh, S. Radha, S. Sakthinathan, T.-W. Chiu and C. Karuppiah. Silver nanoparticles using Cassia Alata and its catalytic reduction activities of Rhodamine6G, Methyl orange and methylene blue dyes, *Inorg. Chem. Commun.*, 2023, **155**, 110985.

[39] S. N. Basahel, T. T. Ali, M. Mokhtar and K. Narasimharao. Influence of crystal structure of nanosized ZrO_2 on photocatalytic degradation of methyl orange, *Nanoscale Res. Lett.*, 2015, **10**, 73.

[40] M. M. M. Mostafa, K. N. Rao, H. S. Harun, S. N. Basahel and I. H. A. El-Maksod. Synthesis and characterization of partially crystalline nanosized ZSM-5 zeolites, *Ceram. Int.*, 2013, **39**, 683–689.

[41] J. F. Gomes, A. C. Garcia, L. H. S. Gasparotto, N. E. de Souza, E. B. Ferreira, C. Pires and G. Tremiliosi-Filho. Influence of silver on the glycerol electro-oxidation over AuAg/C catalysts in alkaline medium: A cyclic voltammetry and in situ FTIR spectroscopy study, *Electrochim. Acta.*, 2014, **144**, 361–368.

[42] I. Capel Berdiell, G. B. Braghin, T. Cordero-Lanzac, P. Beato, L. F. Lundegaard, D. Wragg, S. Bordiga and S. Svelle. Tracking structural deactivation of H-Ferrierite zeolite catalyst during MTH with XRD, *Top. Catal.*, 2023, **66**, 1418–1426.

[43] J. Yuan, J.-J. Zhang, M.-P. Yang, W.-J. Meng, H. Wang and J.-X. Lu, CuO Nanoparticles Supported on TiO_2 with High Efficiency for CO_2 Electrochemical Reduction to Ethanol, *Catalysts*, 2018, **8**, 171.

[44] G. K. Alqurashi, A. Al-Shehri and K. Narasimharao. Effect of TiO_2 morphology on the benzyl alcohol oxidation activity of Fe_2O_3–TiO_2 nanomaterials, *RSC Adv.*, 2016, **6**, 71076–71091.

[45] A. Alshehri and K. Narasimharao, Antimony Substituted Ammonium 12-Molybdophosphoric Acid Catalysts for Gas Phase Chlorobenzene Oxidation, *Catal. Letters*, 2021, **151**, 1025–1037.

[46] A. A. Alshehri and K. Narasimharao, Low Temperature Oxidation of Carbon Monoxide over Mesoporous Au-Fe_2O_3 Catalysts, *J. Nanomater.*, 2017, **2017**, 8707289.

[47] S. N. Basahel, M. Mokhtar, T. T. Ali and K. Narasimharao. Porous Fe_2O_3-ZrO_2 and NiO-ZrO_2 nanocomposites for catalytic N2O decomposition, *Catal. Today.*, 2020, **348**, 166–176.

[48] J. M. Thomas and C. Ducati,Transmission electron microscopy: in *Characterization of Solid Materials and Heterogeneous Catalysts*, Weinheim, 2012, pp. 655–701.

[49] S. N. Basahel, M. Mokhtar, E. Alsharaeh, T. ALI, H. A. Mahmoud and K. Narasimharao. Photocatalytic degradation of *p*-nitrophenol in aqueous suspension by using graphene/ZrO_2 catalysts, *Nanosci. Nanotechnol. Lett.*, 2016, **8**, 448–457.

[50] H. Qu, B. Li and Z. Lv. ZSM-5 templated porous carbon: Synthesis and characterization, *J. Phys. Conf. Ser.*, 2023, **2587**, 12086.

[51] S. Brunauer, L. S. Deming, W. E. Deming and E. Teller. On a theory of the van der waals adsorption of gases, *J. Am. Chem. Soc.*, 1940, **62**, 1723–1732.

[52] P. Llewellyn, E. Bloch and S. Bourrelly, Surface area/porosity, adsorption, diffusion: in *Characterization of Solid Materials and Heterogeneous Catalysis*, eds. M. Che and J. Védrine, Wiley-VCH Verlag GmbH & Co, Weinheim, 2015, pp. 853–879.

[53] N. Albeladi, Q. A. Alsulami and K. Narasimharao. Zn deposited nickel silicate nanotubes as efficient CO2 methanation catalysts, *Mol. Catal.*, 2024, **556**, 113949.

[54] A. Pintar, J. Batista and S. Hočevar. TPR, TPO, and TPD examinations of Cu0.15Ce0.85O2−y mixed oxides prepared by co-precipitation, by the sol–gel peroxide route, and by citric acid-assisted synthesis, *J. Colloid Interface Sci.*, 2005, **285**, 218–231.

[55] K. N. Rao, L. D. Dingwall, P. L. Gai, A. F. Lee, S. J. Tavener, N. A. Young and K. Wilson. Synthesis and characterization of nanoporous phospho-tungstate organic–inorganic hybrid materials, *J. Mater. Chem.*, 2008, **18**, 868–874.

[56] L. Li, X. Cui, J. Li and J. Wang. Synthesis of SAPO-34/ZSM-5 composite and its catalytic performance in the conversion of methanol to hydrocarbons, *J. Braz. Chem. Soc.*, 2015, **26**, 290–296.

[57] M. Mokhtar, B. F. A. Alhashedi, H. A. Kashmery, N. S. Ahmed, T. S. Saleh and K. Narasimharao. Highly efficient nanosized mesoporous cumgal ternary oxide catalyst for nitro-alcohol synthesis: Ultrasound-assisted sustainable green perspective for the Henry reaction, *ACS Omega.*, 2020, **5**, 6532–6544.

[58] T. T. Ali, K. Narasimharao, N. S. Ahmed, S. Basahel, S. Al-Thabaiti and M. Mokhtar. Nanosized iron and nickel oxide zirconia supported catalysts for benzylation of benzene: Role of metal oxide support interaction, *Appl. Catal. A Gen.*, 2014, **486**, 19–31.

Chapter 4
Green synthesis methods

4.1 Introduction

Inorganic catalysts, also known as heterogeneous catalysts, typically encompass metals and metal oxides dispersed on porous materials, mimicking the intricate functionality of nature's catalysts (enzymes). Reactants adsorb onto the active sites on the metal surface during the catalytic process. Generally, heterogeneous catalysts prefer inorganic materials consisting of metals and metal oxides, given their high thermal stability, which is crucial for various industrial applications. Researchers are interested in synthesizing heterogeneous catalysts in nanomaterials (NMs) for various reasons. One of the primary reasons is to improve the catalytic activity and efficiency of the catalyst. NMs, such as metal nanoparticles (NPs) or metal-based oxides, have unique physicochemical properties compared to their bulk counterparts, such as larger surface area, higher reactivity, and better dispersion, which can significantly enhance the catalytic performance. Additionally, the use of NMs as catalysts allows for better control over the catalytic process including reaction conditions, product selectivity, and recyclability. The NM production typically involves two primary approaches: top-down and bottom-up techniques. Top-down methods include physical and chemical techniques. Bottom-up methods, on the other hand, include chemical and biological methods (details in Chapter II). The latter, biological method, involves the use of living organisms or biological systems to synthesize NMs. Biological methods, similar to chemical means, involving the use of a reducing agent to reduce a precursor containing metallic ions, causing the metallic ions to transform into NPs. The primary reason why the biosynthetic method is considered superior to physical and chemical methods for the production of metallic NPs is its ability to achieve excellent monodispersity and stability. Due to the presence of naturally occurring reducing and stabilizing materials, the biosynthetic method not only produces highly uniform metallic NPs but also minimizes the use of toxic chemicals and byproducts, making it more environmentally friendly, and simple method.

4.2 Biological (green) methods

Biological methods involve utilizing organisms, such as plants, enzymes, microbes, proteins, and extracts, to supply electrons to reduce precursor mixtures containing metal ions, consequently transforming metal ions into NMs and stabilizing NMs. Biomaterial methods can be classified based on the type of organisms that generate electrons as the reducing agents. The two most commonly employed biological materials are **(i)** plants and **(ii)** the microscopic realm of microorganisms (see Figure 4.1). Indeed, the

https://doi.org/10.1515/9783111316819-004

primary advantage of biological methods over chemical and physical approaches lies in their sustainability, as microorganisms such as bacteria can replicate exponentially and sustain the system autonomously. Similarly, plants can be replanted and cultivated for long-term environmental compatibility. Moreover, this method offers a viable option for the mass production of NMs due to its readily available, flexible, eco-friendly nature, scalability, and the capacity of its biological components to serve as both reducing and capping agents, thereby conserving energy by eliminating the need for high temperatures and pressures. This approach also possesses a cost-effective advantage. While biological methods for producing NMs offer certain advantages, they may pose limitations regarding the size range and range of metal combinations that can be achieved. Additionally, some plant extracts or microorganisms may have limited industrial-level applicability, as they only effectively reduce a relatively small number of metals to form NMs. This is because each material requires a specific production setup.

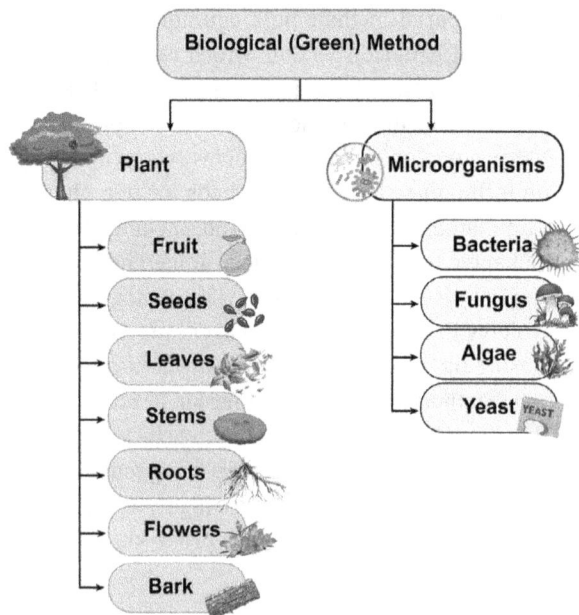

Figure 4.1: Different methods for using green synthesis of nanomaterials.

4.3 Plant-mediated method

There are distinct mechanisms for forming NMs using different plants or from different parts of plant extracts. Plant extracts commonly derive from (i) leaves, (ii) roots, (iii) flowers, (iv) fruits, (v) seeds, (vi) stems, and (vii) bark; these all function as stabilizers, coverings, oxidizing, and reducing agents in the process [1]. It is fascinating to consider that extract solutions may contain proteins, enzymes, vitamins, polysaccharides, carbohydrates, organic acids, and amino acids. In this scenario, these phytochemicals act as stabilizing agents, as external stabilization is limited [2]. Various extraction techniques have been documented, encompassing methods such as maceration, Soxhlet, microwave, supercritical fluid, pulse electric field, pressure liquid, ultrasound, and enzyme-based approaches. It is crucial to select the most appropriate extraction technique and solvent for generating biomolecules. The maceration method, a simple technique, is commonly employed in synthesizing NMs using different plant extracts [3]. This process involves drying the plant for an extended period, grinding it into smaller pieces, placing it in a solvent, and subjecting it to periodic vibration. The resultant mixture is then filtered to isolate the extract, as depicted in Figure 4.2. The significant parameters governing the efficiency of the extraction process depend on the specific plant material and type of solvent being used such as water, oil, or alcohol. Following this, the extract solution is combined with a solution containing metal salts. The mixture is agitated continuously for several hours at room temperature, ultimately resulting in the formation of solid NPs, as illustrated in Figure 4.2.

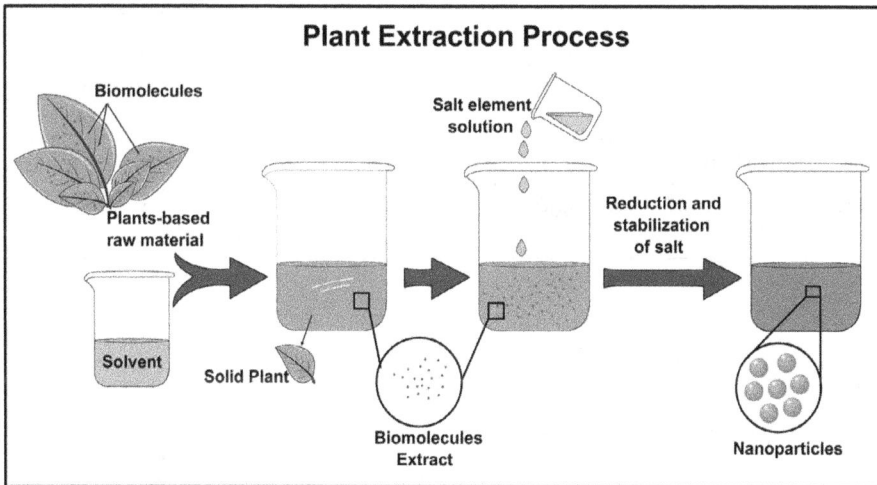

Figure 4.2: A schematic representation of the extraction using the Maceration method (Adapted from ref. [4]).

There are several examples of the reduction of metals, such as silver ions (Ag^+) from an aqueous solution of silver nitrate ($AgNO_3$) using plant extracts, such as extracts from *Clitoria ternatea* [5], the green tea (*Camellia sinensis*) [6], alfalfa (*Medicago sativa*) [7], and lemongrass (*Cymbopogon citratus* L.) [8], more in Table 4.1. Also, reduction of gold ions (Au^+) from an aqueous solution of chloroauric acid ($HauCl_4$) using plant extracts, such as extracts from Cinnamon bark [9], *Limnophila rugosa* leaves

[10], and *Cistanche deserticola* [11], etc. In addition, metal oxides such as titanium dioxide (TiO_2) NPs can also synthesized using precursor of titanium tetrachloride ($TiCl_4$) and plant extracts such as Papaya (Caricaceae) [12], *Syzygium cumini* leaves [13], *Calotropis gigantea* extract [14], and Bengal gram beans (*Cicer arietinum* L.) [15]. Zinc oxide (ZnO) also used precursor zinc nitrate hexahydrate [$Zn(NO_3)_2.6H_2O$] and plant extracts such as *Phoenix dactylifera* waste [16], *Artocarpus hetrophyllus* leaves [17], and Areca nut (*Areca catechu*) [18]. Another metal oxide Fe_xO_x using iron(III) chloride hexahydrate ($FeCl_3.6H_2O$) as precursor and plant extracts such as *Carica papaya* [19], *Psidium guavaja-Moringa oleifera* [20], and Amla seed [21]. While NiO NPs use precursor nickel (II) nitrate hexahydrate [$Ni(NO_3)_2.6H_2O$] mixed with plant extracts such as Olive tree leaves [22], *Syzygium cumini* plant [23], and fresh tea leaves (*Camellia sinensis* plant) [24].

Table 4.1: Different types of nanomaterial synthesized using the plant extract.

Plant part	Plant name	Metal salt	Particle types	Av. size (nm)	Applications
Plant	*Clitoria ternatea*	$AgNO_3$	Ag NPs	–	Photodegradation of Methylene blue [5]
leaves	Green tea (*Camellia sinensis*)	$AgNO_3$	Ag NPs	25	Photodegradation of Methylene blue and antibacterial activity [6]
Plant	Alfalfa (*Medicago sativa*)	$AgNO_3$	Ag NPs	12	Anti–bacterial [7]
leaves	Lemongrass (*Cymbopogon citratus* L.)	$AgNO_3$	Ag NPs	332	Biopesticides for lichen inhibition on stones [8]
Roots	Sweet potatoes (*Ipomoea batatas* L. var. Rancing)	$AgNO_3$	Ag NPs	26.44	Potential filler for 3D-printed electroactive and anti-infection scaffolds [25]
Bark	Cinnamon bark [9]	$HauCl_4$	Au NPs	5–35	Quencher of eosin Y dye [9]
leaves	*Limnophila rugosa* leaves	$HauCl_4$	Au NPs	122.8	Catalytic activity in reduction of nitrophenols [10]
Plant	Cistanche deserticola	$HauCl_4$	Au NPs	122.8	Catalytic activity in reduction of nitrophenols [11]
Fruit	Papaya (Caricaceae)	$TiCl_4$	TiO_2 NPs	–	Antifungal application [12]
leaves	*Syzygium cumini* leaves	$TiCl_4$	TiO_2 NPs	12–13	Photocatalytic activity and nematicidal evaluation [13]
Flower	*Calotropis gigantea* extract [14]	$TiCl_4$	TiO_2 NPs	32.34	Photocatalytic and photovoltaic application [14]

Table 4.1 (continued)

Plant part	Plant name	Metal salt	Particle types	Av. size (nm)	Applications
Beans	Bengal gram beans (*Cicer arietinum* L.)	$TiCl_4$	TiO_2 NPs	14	Lithium ion battery application [15]
Plane	*Phoenix dactylifera* waste	$Zn(NO_3)_2.6H_2O$	ZnO NPs	31.6	Dye degradation and antibacterial of wastewater [16]
Leaves	*Artocarpus hetrophyllus* leaves	$Zn(NO_3)_2.6H_2O$	ZnO NPs	–	Degradation of Congo red dye in water [17]
Fruit	Areca nut (*Areca catechu*)	$Zn(NO_3)_2.6H_2O$	ZnO NPs	5–30	Degradation of dye, polyethylene, and antibacterial performance in waste water treatment [18]
Leaf	*Carica papaya* leaf extract	$FeCl_3.6H_2O$	α-Fe_2O_3 NPs	21.59	Photocatalytic degradation of Remazol yellow RR dye and antibacterial activity [19]
Fruit	*Psidium guavaja-Moringa oleifera*	$FeCl_3.6H_2O$	FeO NPs	75–82	Antibacterial and photocatalytic degradation of Methylene blue dye [20]
Seed	Amla seed	$FeCl_3$	FeO NPs	2–5	Photocatalytic degradation of Methylene blue dye [21]
Leaves	Olive tree leaves	$Ni(NO_3)_2.6H_2O$	NiO NPs	42	Adsorption of dye [22]
Fruit	*Syzygium cumini* plant	$Ni(NO_3)_2.6H_2O$	NiO NPs	10.4	Antioxidant and dyes removal [23]
Leaves	Fresh tea leaves (*Camellia sinensis* plant)	$Ni(NO_3)_2.6H_2O$	NiO NPs	3–5	Photocatalytic degradation of both cationic, anionic dyes, and their mixture [24]

To further reduce environmental pollution and lower the cost of synthesis, efforts have been made to synthesize zeolite through green pathways such as solvent-free during hydrothermal or the use of sustainable precursors and mother-liquor recycling. In addition, some researchers have also explored the use of free templates and other green methods to further minimize environmental impact [26]. In the synthesis of zeolite, the seed-directed method provides an alternate approach to using organic templates. This technique can function in two distinct ways: (i) The seed acts as an inorganic nucleus that can cause the surrounding species to form different topologies, distinct from those of the seed or (ii) The seed can also function as an essential building unit that allows direct crystallization, following the topology of the seed without needing an organic template [27]. Furthermore, in the search for sustainable and eco-friendly methods of nanozeolite synthesis, researchers have identified several potential bio-silica sources as alternatives to traditional synthetic materials. For example, recent studies have reported the use of coal fly ash (industrial waste) [28–34], iron ore tailing [35], rice husk ash [36–41], sugarcane bagasse ash [42], bauxite [43], and bamboo leaf biomass [44] as renewable precursors. Bio-aluminum precursors have also been reported to use red mud as a source for synthesizing magnetic zeolite [44]. In addition to these bio-precursors, alumina and silica precursors such as natural kaolin [45–53] and perlite [54] can also be utilized. Furthermore, alternative methods that are solvent-free to reduce or eliminate mother liquors, which can have harmful effects on the environment, are being implemented. Interestingly, the main focus in the development of the synthesis of zeolite using environmental approaches has been on three key parameters: the precursor, the free assistance of organic structure, and the free solvent. For example, the use of raw kaolin as an aluminum source to produce Zeolite A for the removal of Cr(III) from tannery wastewater. The kaolin undergoes calcination at 600 °C for 3 h to form metakaolin. This is then treated with (3 M) NaOH and stirred for 1 h at 50 °C until a gel forms. The gel ages at room temperature for 24 h before being placed in a water bath at 100 °C for 3 h. This process results in the formation of Zeolite A with a molar composition of 1 Na_2O:1 Al_2O_3:2 SiO_2:37 H_2O. After the bottle was cooled, the material was washed in distilled water until the pH reached less than 10 and was then dried at 80 °C overnight. Next, 3 M NaOH was added without undergoing aging under hydrothermal crystallization, resulting in the production of Zeolite A. This crystalline material took on a cubic shape with an average particle size of 5 μm [45, 46]. Zeolite A's removal capacity for Cr(III) from tannery wastewater was found to be 99.8% [45].

4.4 Microorganism method

Microorganisms encompass two essential reduction materials: (i) nicotinamide adenine dinucleotide (NADH) and (ii) NADH-dependent nitrate reductase, which are employed specifically for silver (Ag) and gold (Au) ions. Nevertheless, the bioreduction mechanism for transforming metal salt ions into metallic NMs remains largely unexplored.

4.4.1 Bacteria-mediate synthesis

Bacteria can serve as effective agents for synthesizing a variety of new NMs. Two primary approaches for leveraging bacteria in NM production exist the extracellular and intracellular methods. The extracellular method is considered advantageous over the intracellular method as it does not require complex procedures to extract NMs from the bacteria and offers a shorter synthesis time. Bacteria possess the advantage of rapid reproduction, which makes them highly attractive for mass-producing NPs in industrial settings. Furthermore, bacterial cells are cost-effective, can be frozen for storage and regrown as needed, and facilitate sustainable manufacturing processes. Biomasses of various types can produce identical NPs, offering flexibility and avoiding dependence on specific bacterial species. Additionally, a single species of bacteria has the potential to produce multiple types of NPs, unlike chemical synthesis, which necessitates distinct chemicals for each reaction type. A limitation of bacterial-based synthesis is the necessity to separate the biomass and NPs after synthesis, which can be cumbersome and costly in terms of time and resources. The selection of a bacterial species for NM production depends on various factors, such as cost, eco-friendly production methods, and scalability of the synthesis process. The literature has demonstrated that a range of bacteria, including *Actinobacter* spp., *Escherichia coli*, *Klebsiella pneumonia*, *Lactobacillus* spp., *Bacillus cereus*, *Corynebacterium* spp., *Pseudomonas* spp., and *Enterobacter cloacae*, are capable of reducing metallic ions to form NPs. Bacteria can produce NPs within their cells (intercellular synthesis) or outside (extracellular synthesis). The latter approach is more commonly employed due to its simplicity compared to the former. However, a drawback of both methods is separating the biomass from the synthesized NPs, which can be time-consuming and resource-intensive.

For example, Jiang et al. [55] synthesized the bimetallic Au–Ag NPs using Bacteria called *Escherichia coli* (Figure 4.3a). They used 100 μL of a 10 mM aqueous solution of $HAuCl_4$ and 600 μL of an *E. coli* culture, with a cell count of 1×10^8 cells per milliliter (mL), were mixed thoroughly in an ice bath while stirring vigorously. Two milliliters of normal saline was then added, followed by the addition of sodium hydroxide (0.5 M) to increase the solution's pH value to greater than 12. After a 5-min reaction time in the alkaline environment, the remaining *E. coli* were separated from the solution via centrifugation (10,000 g). Initially, 30 min of continuous production of gold NPs took place in the supernatant liquor. When it was finished, 100 μL of a 10 mM silver nitrate ($AgNO_3$) solution was added dropwise to the mixture while stirring for 1 min. The mixture was then kept in the dark for 12 h to facilitate the generation of bimetallic gold-silver NPs. A dialysis bag with a 3,000 Da molecular weight cutoff was used to remove unreacted ions, including Au^+, Ag^+, OH^-, and Cl^-, using double-distilled water for 48 h. The TEM image (Figure 4.3b) showed homogenous and rounded bimetallic Au-Ag particles (14–18 nm). Another attempt by Kouhkan et al. [56] used *Lactobacillus casei* (Figure 4.3c) for biosynthesized copper oxide nanoparticles (CuO NPs). A culture of *L. casei* was prepared in a round-bottom flask containing Mannitol-Salt Agar (MRS) broth and shaken on a rotary shaker at 200 rpm at a temperature of 37 °C. The pH of the broth culture was then adjusted to 6 using 0.8 M NaOH, which delayed the transformation process. A solution of copper sulfate (1 mM) was then added to the mixture, and the entire solution was incubated for 48 h at 37 °C. The medium changed color from yellow to dark brown at the bottom of the flask, indicating the formation of copper NPs. The product was filtered and rinsed with deionized water and then dried in a hot air oven at 40 °C for 4 h.

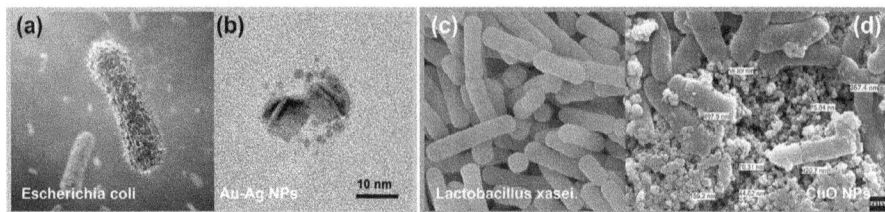

Figure 4.3: (a) Bacteria *Escherichia coli* [57]; **(b)** TEM image of bimetallic Au-Ag nanoparticle (adapted from ref. [55]); **(c)** bacteria *Lactobacillus casei*; and **(d)** SEM image of CuO NPs (adapted from ref. [56]).

The SEM of CuO NPs is synthesized with *L. casei* (Figure 4.3d). It is spherical in shape and uniform in size from 40 to 110 nm (see Table 4.2).

4.4.2 Fungi-mediated synthesis

Fungi have a significant impact on addressing major global challenges related to sustainability. They efficiently enhance resource utilization by converting waste into valuable food and feed ingredients, making crop plants resilient to climate change, and serving as hosts for producing innovative biological medications [58]. Fungal synthesis of NPs represents an attractive approach due to its eco-friendliness and the capability to yield a substantial amount of stable NPs with minimal toxicity. This synthesis process involves employing fungi as reducing agents, wherein they produce extracellular proteins that can stabilize the NPs and render them biocompatible. By manipulating culture conditions like ion precursor concentration, temperature, pH, and biomass quantity, a variety of fungi species can be utilized to synthesize NPs with varying characteristics such as size and surface charge [59]. The biogenic synthesis of NPs from fungi can occur either **(i)** intracellularly or **(ii)** extracellularly, with the latter being the most commonly employed method. Intracellular synthesis happens within fungal cells and necessitates intricate extraction techniques after NP formation. It involves capturing and transporting metal ions within the microbial cells via electrostatic interactions between the ions and the negatively charged cell wall enzymes. There are a variety of possible mechanisms for the formation of NMs by intracellular reductase: (i) periplasmic reductase can directly reduce metal ions (M^+) into metallic form (M), (ii) bio-reduction within the cytoplasm or periplasm can transform M^+ into M, or (iii) bioconversion of M^{2+} to M^+ within the cytosol facilitates the formation of M. Intracellular enzymes, such as cytochrome oxidases, facilitate the reduction of metal ions to NMs through the transfer of electrons between substances present within the cytosol such as the coenzymes nicotinamide adenine dinucleotide (NAD^+) and nicotinamide adenine dinucleotide phosphate ($NADP^+$), vitamins, and organic acids. With the extracellular method, metal ions adhere to the surface of bacterial or fungal cells and are reduced in the presence of enzymes and proteins as well as cell wall components and organic molecules in the culture medium.

Through extracellular synthesis, the detained metal ions within the cells undergo bio-reduction, giving rise to metallic nuclei, which expand in size [60]. Although it does not require extracting NPs from the cell and does not require NP separation, it still requires purifying the NP dispersion to eradicate fungal remnants and impurities. Similar to plant– and bacterial-based NP synthesis, all biosynthesized NPs require additional purification steps to ensure their purity. Extracellular enzymes, such as nitrate reductase, can facilitate the transfer of electrons from hydroxyl groups (OH^-) to metal ions. Microbial proteins, which possess functional groups such as ($-NH_2$), ($-OH$), ($-SH$), or ($-COOH$), assist in stabilizing NMs by providing binding sites to metal ions. This mechanism reduces metallic ions into NMs on the cell wall or in the periplasmic space. Furthermore, in some cases, proteins serve as both the primary reducing and capping agents during the formation of NMs, and they continue to stabilize the NMs even after their creation. Fungi possess several advantages over other microorganisms – they exhibit enhanced tolerance to heavy metals, make handling easier, and produce substantial biomass. Furthermore, they offer excellent potential in producing bioactive substances that can be harnessed for various applications. In summary, the utilization of fungi for biogenic synthesis of NPs holds great promise and presents exciting opportunities for further research in health and agriculture. Research has shown that fungi can produce mono-dispersed NMs in diverse chemical compositions and particle sizes. When contrasted with other microorganisms, such as bacteria, fungi possess additional characteristics contributing to synthesizing metallic NMs. Notably, fungi secrete substantial quantities of enzymes and proteins per unit mass, resulting in a higher yield of NMs than in bacterial synthesis. This process can occur through extracellular and intracellular pathways for synthesizing gold (Au) and silver (Ag) nanoparticles. Various fungal species, including *Penicillium* spp., *Fusarium* spp., *Fusarium oxysporum*, *F. semitectum*, *F. acuminatum*, *F. solani*, *Cladosporium cladosporioides*, *T. asperellum*, and *Aspergillus* spp.

Qu et al.'s [61] work illustrates the use of fungi in the biosynthesis of gold nanoparticles. Their research demonstrates that a cell-free extract derived from the fungus *Trichoderma* sp (Figure 4.4a). (WL-Go) enables the production of gold nanoparticles see Figure 4.4b. Strain WL-Go was cultivated in the medium known as MMB until it reached the stationary phase in an aerated environment. The fungal cells were then separated from the culture using qualitative filter papers and rinsed thoroughly with sterile ultrapure water three times. To prepare the cell-free extracts, the cells of strain WL-Go were suspended in a 50 mmol/L phosphate sodium buffer at pH 7.0 and then lysed using ultrasonication (20 kHz) for 30 min. Cell-free extracts with varying protein concentrations of 22.5, 45, 90, 180, and 360 mg/L were prepared by adjusting the extract with a phosphate sodium buffer. The impact of these varying concentrations on the biosynthesis of AuNPs was then examined. The concentration of $HAuCl_4$ used in the mixture was set to 1 mmol/L. Additionally, Murillo-Rabago et al. [62] utilized the same fungus, *Trichoderma* sp., to synthesize silver (Ag) nanoparticles by employing both intracellular and extracellular approaches. To isolate the intracellular components of the fungal biomass, the biomass was first separated from the culture broth by washing with distilled water, then homogenized in an agate mortar with deionized water at a ratio of 1:1 (g/mL), and finally, the resulting homogeneous mixture was subject to centrifugation at 10,000 rpm for 15 min. The pellet was discarded, and the resulting aqueous extract was further filtered using a 0.22 μm nitrocellulose membrane to remove any remaining biomass residue. While extracellular components of fungal strains were obtained by washing the biomass with deionized water, weighing it, and placing it in a flask with 100 mL of deionized water. The cultures were maintained at 26 °C for 3 days in a shaker set at 125 rpm. Subsequently, the biomass was removed through filtration, and the resulting supernatant was centrifuged and filtered with a 0.22 μm nitrocellulose membrane. The optimized protocol for synthesizing AgNPs involved using supernatants and extracts from *T. harzianum*. To obtain the supernatants, 10 g of previously washed biomass was incubated in 100 mL of deionized water for 3 days. The formation of NPs was most effective with a ratio of 1:3 v/v of (extract or supernatant)/1 mM $AgNO_3$ and incubation for 3 days at 60 °C. The Ag NPs obtained using the supernatant from *T. harzianum* (AgNPs-TS) exhibited an average size of 9.6 nm, with a range from 1 to 33 nm (Figure 4.4c). In contrast, AgNPs obtained with the extract from *T. harzianum* (AgNPs-TE) had an average size of 19.1 nm and a size range from 3 to 59 nm (Figure 4.4d).

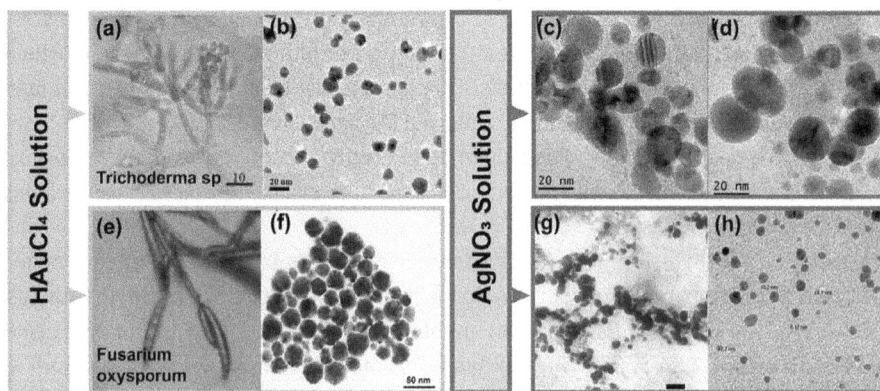

Figure 4.4: **(a)** Fungi *Trichoderma harzianum* (adapted from ref. [63]); **(b)** TEM Au nanoparticle (adapted from ref. [61]); **(c)** TEM Ag nanoparticle synthesized using the supernatant (extracellular) of *T. harzianum*, and **(d)** TEM Ag nanoparticle synthesized using the extract of *T. harzianum* (adapted from ref. [62]); **(e)** fungi *Fusarium* spp. (adapted from ref. [64]); **(f)** TEM Au nanoparticle synthesized using fungi *Fusarium* spp. Adapted from ref. [65]; **(g)** TEM Ag nanoparticle synthesized using fungi *Fusarium* spp. (adapted from ref. [66]); **(h)** TEM Ag nanoparticle synthesized using fungi *Fusarium* spp. (adapted from ref. [67]).

Naimi-Shamel et al. [65] synthesized gold (Au) nanoparticles using fungus called *F. oxysporum*. (Figure 4.4e) hyphae were inoculated into Sabouraud dextrose broth medium and then cultivated in conical flasks at a temperature of 30 °C with an agitation rate of 200 rpm for 72 h. The culture was subsequently centrifuged at 6,000 rpm for 10 min, and the resultant supernatant was utilized for GNPs synthesis. A solution of 1 mmol/L $HAuCl_4$ (1 mL) was mixed with 1 mL of the culture supernatant, resulting in a final concentration of 1 mmol/L $HAuCl_4$. After incubation at 30 °C with constant agitation (150 g) for a duration of 24 h the formation of GNPs was visually confirmed through a noticeable change in the color of the liquid from clear to pink or purple, indicating the reduction of HA. Ahmed et al. [66] utilized *F. oxysporum* to synthesize Ag NPs. To synthesize Ag NPs with *F. oxysporum*, 10 g of fungal biomass was added to 100 mL of distilled water in a conical flask. $AgNO_3$ was added to the solution to obtain a final concentration of 10^{-3} M. The reaction was carried out in the dark, and periodic aliquots of the reaction solution were removed for UV-vis and fluorescence spectroscopic measurements. Also, Husseiny et al. [67] used *F. oxysporum* to obtain Ag NPs. *F. oxysporum* was grown in PDA broth medium in flasks at 28 °C for 5 days. The fungal biomass was then separated from the medium by filtering through Whatman filter paper number 1 and washed three times with sterile distilled water to remove any residual nutrients. An $AgNO_3$ solution was added to achieve an overall silver concentration of 1 mM (0.017/100 mL), and the mixture was left in the incubator at 28 °C for 5 days. The interactions were carried out in the dark, with a control group containing only the metal ions without fungal.

Table 4.2: Different types of nanomaterial synthesized using the microorganism.

Micro.[a]	Species names	Metal precursor	Particle types	Av. size (nm)[b]	Applications
Bacteria	*Escherichia coli*	$HAuCl_4 + AgNO_3$	Au-Ag NPs	14–18	Biomedical: ultrafast colorimetric detection of H_2O_2, photothermal, and antibiotic therapy [55]
Bacteria	*Lactobacillus casei*	$CuSO_4$	CuO NPs	40–110	Anticancer and antibacterial activities [56]
Fungi	*Trichoderma harzianum* (WL-Go)	$HAuCl_4$	Au NPs	20–30	Catalytic activity on nitro-aromatics reduction and azo dyes decolorization [61]
Fungi	*T. harzianum* (supernatant-extracellular)	$AgNO_3$	Ag NPs	9.6	Bacterial inhibition [62]
	T. harzianum (extract-intracellular)	$AgNO_3$	Ag NPs	19.1	
Fungi	*Fusarium oxysporum*	$HAuCl_4$	Au NPs	22–30	Antibacterial activity of tetracycline conjugant [65]

Table 4.2 (continued)

Micro.[a]	Species names	Metal precursor	Particle types	Av. size (nm)[b]	Applications
Fungi	*F. oxysporum* (extracellular)	AgNO$_3$	Ag NPs	5–50	– [66]
Fungi	*F. oxysporum*	AgNO$_3$	Ag NPs	5–13	Antibacterial and antitumor activities [67]
Algae	Chorella extract	Zinc acetate	ZnO NPs	20	Photocatalytic activity for degradation of dibenzothiophene (DBT) [68]
Algae	*Phaeodactylum tricornutum*	(CH$_3$CH(O-)CO$_2$NH$_4$)$_2$Ti(OH)$_2$	Ti NPs	300–500	Antimicrobial, biogenic and antistatic studies [69]
Yeast	*Saccharomyces cerevisiae*	Zinc acetate	ZnO	15	Antibacterial activity and photocatalytic degradation [70]

[a]Microorganism.
[b]Average size (nm).

4.4.3 Algae-mediated synthesis

Microalgae, being aquatic, photosynthetic microorganisms, can be either unicellular (e.g., chlorella) or multicellular (e.g., brown algae) [71]. Algae have been found to play a considerable role in the biological synthesis of NMs and the accumulation of various heavy metals comparable to other microorganisms [69, 70]. Algae have proven to be effective in synthesizing ZnO nanoparticles and the large-scale production of gold and silver nanoparticles. Notably, microalgae are recognized for their ability to transform hazardous forms of metals into their harmless counterparts. The biological compounds in microalgae can effectively enhance specific properties of ZnO nanoparticles, resulting in expanded application possibilities. Various compounds such as phenols, tannins, alkaloids, sterols, saponins, flavonoids, and tocopherols are commonly found in these microalgae. Various microalgae strains, including *Phaeodactylum tricornutum, Sargassum muticum, Sargassum myriocystum, Chlorella vulgaris, Padina* spp, and other in Figure 4.5.

For example, Khalafi et al. [68] synthesized ZnO using microalgae *Chlorella* extract (Figure 4.6a). The soaking method was commonly used when extracting Chlorella from microalgae for experimental purposes. This process involved mixing 1 g of Chlorella powder with 200 mL of deionized (DI) water in a 250 mL Erlenmeyer flask, followed by vigorously magnetic stirring at 80 °C for 15 min while using a stirrer heater. The heated Chlorella extract was subsequently filtered through a Millipore filter with a pore size of 0.1 μm, and the filtered liquid was preserved at 4 °C for further experiments. An 80 mL

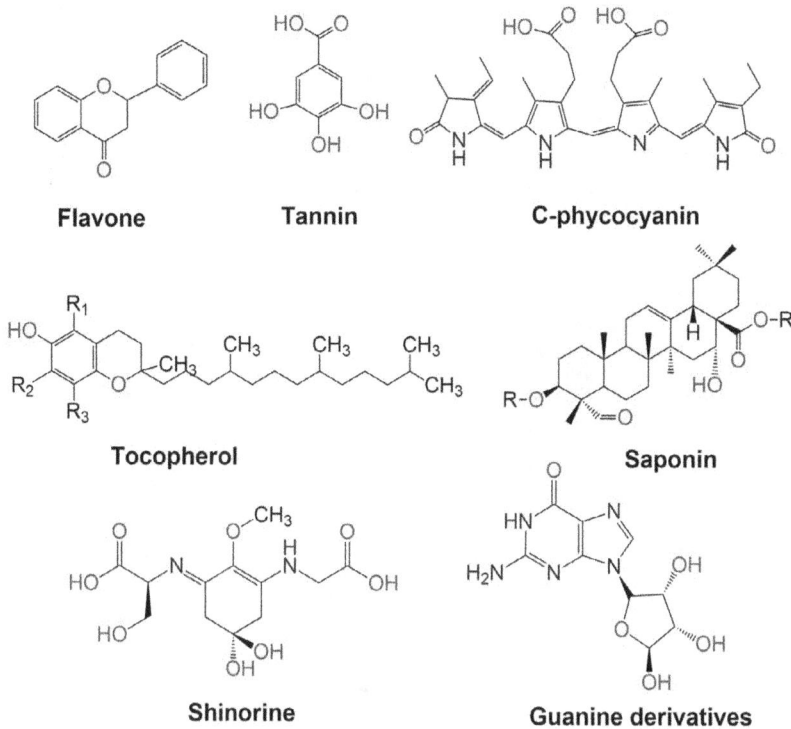

Figure 4.5: Several biological compounds found in microalgae extracts (adapted from ref. [70]).

solution of zinc acetate was combined with 20 mL of clean algal extract liquid at a temperature of 58 °C and agitated continuously for 60 min at a speed of 150 rpm. Heating was necessary to accelerate the reduction process, facilitated by the electron-rich organic compounds in microalgae Chlorella. The primary pH of the mixture solution was initially adjusted at a level of 8, but it was then decreased to 5.5 by the end of the reaction due to a reduction in alkaline OH groups. The milky-looking solution was further stirred for an additional 25 min at a temperature of 85 °C. The obtained precipitate was then centrifuged and rinsed with deionized water, followed by drying at a temperature of 50 °C. TEM analysis revealed that the ZnO nanoparticles obtained were shaped in hexagonal patterns with a size of approximately 20 nm (Figure 4.6b). In a related study, Caliskan et al. [69] successfully synthesized titanium nanoparticles using microalgae called *P. tricornutum*, see Figure 4.6c. This particular marine microalga, *P. tricornutum*, was acquired from the Culture Collection of Algae (SAG). Its cultivation was maintained through regular subculturing methods under laboratory conditions at a temperature of 24 ± 1 °C and an initial pH value of between 8 ± 0.5. These microalgae were cultured in the F/2 medium, which was enriched with 15 g/L of sea salt. Phototrophic batch cultures were grown in 250-mL conical flasks containing 100 mL of the liquid medium. The flasks were then incubated in an orbital shaking incubator, operating at 200 rpm.

The synthesis of titanium nanoparticles was conducted using titanium (IV) bis (ammonium lactate) dihydroxide [$(CH_3CH(O-)CO_2NH_4)_2Ti(OH)_2$] dissolved in solutions and *P. tricornutum* microalgae. Initially, stock solutions of titanium in different concentrations ranging from 0.1 to 10 mM were prepared for nanoparticle production. Additionally, different ratios between the titanium solution and the supernatant volume (0.2–10 mL per volume: volume or v:v) were created separately. The obtained material has size of 300–500 nm as shown in Figure 4.6d.

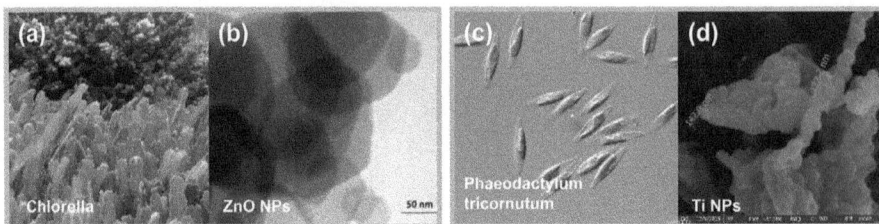

Figure 4.6: (a) Microalgae *chlorella*; **(b)** TEM image of ZnO nanoparticle (adapted from ref. [68]); **(c)** microalgae *Phaeodactylum tricornutum*; and **(d)** SEM image of Ti NPs (adapted from ref. [69]).

4.4.4 Yeast-mediated synthesis

Certain microorganisms known as yeasts, who are single-cell eukaryotes, have evolved from their multicellular ancestors. Studies have demonstrated that yeasts can be used to synthesize nano-materials with remarkable success [70]. Yeasts possess the ability to acclimate to hazardous metals through multiple detoxifying techniques, including bio-sorption, extracellular sequestration, chelation, and bioprecipitation. The advantage of producing yeast in laboratory settings is that its growth can be easily controlled [72]. Additionally, the rapid growth of yeast strains combined with the simplicity of using basic nutrients contributes significantly to the large-scale production of metal nanoparticles. There are approximately 1,500 species of yeast, with many being extensively applied in the production of metallic nano-materials, such as baker's yeast *Saccharomyces cerevisiae, Nematospora coryli, S. pombe*, yeast strain *MKY3, Yarrowia lipolytica* cells, and *Torulopsis* spp. This is because of their expansive surfaces, allowing them to absorb and accumulate substantial quantities of hazardous heavy metals from their surroundings. These mechanisms, which yeast cells have developed, are being utilized to generate nano-materials and increase their durability, leading to differences in particle characteristics and size.

Several studies have been carried out to examine the synthesis of metallic nano-materials utilizing yeast. For instance, El-Khawaga et al. [70] successfully synthesized ZnO NPs using *S. cerevisiae*, commonly known as baking yeast. The process of synthesizing ZnO NPs involved dissolving 2 mM of zinc acetate in 100 mL of doubly distilled water, followed by adding 2 mM of baker's yeast extract. The mixture was then vigorously stirred at a temperature of 50 °C for 12 h, at which time a milky color was observed, indicating the formation of ZnO nanoparticles. Separation was achieved by centrifugation in a solution of ethanol and water, followed by additional drying at a temperature of 100 °C for one night. Finally, the sample heat in a furnace at a temperature of 200 °C for approximately 6 h. SEM image shows (Figure 4.7b) that ZnO NPs have particles of uniform aggregation pattern and possess a spherical morphology with size 15 nm. In another study by Zhang et al. [72], they used *Magnusiomyces ingens* LH-F1 yeast for the synthesis of Ag NPs. The F1 strain LH-F1 was cultivated aerobically at 30 °C in a culture medium until it reached the late logarithmic phase. Cells were then harvested by centrifugation at a rate of 10 g for 10 min at 4 °C, followed by washing twice with double-distilled water and storage at −80 °C before utilization. Later, the cells were re-suspended in the same ddH$_2$O to achieve an optimal density of OD60 (equivalent to a dry cell weight of 2.2 mg/mL). To synthesize Au NPs, a HAuCl$_4$ stock solution (with a concentration of 50 mM) was added to the re-suspended yeast cells, maintaining a final concentration of 1.0 mM. The mixture was then incubated at a temperature of 30 °C for a period of 24 h while being constantly agitated. Following incubation, the mixture was then subjected to centrifugation at a rate of 3,000 g for 5 min, resulting in a supernatant that was filtered using 0.45-μm syringe millipore filters to eliminate any remaining cell debris. The results of these experiments were monitored by observing the UV-vis spectra of the reaction mixtures, which were then used to evaluate the formation of Au NPs. The biomass concentration, as well as the initial gold ion concentration, was seen to significantly influence the synthesis of Au NPs by LH-F1 yeast. The outcome of the synthesis process entailed the formation of spherical, plate-shaped (triangular, hexagonal, and pentagonal), and irregularly shaped nanoparticles, each with an average size of approximately 80 nm (Figure 4.7d).

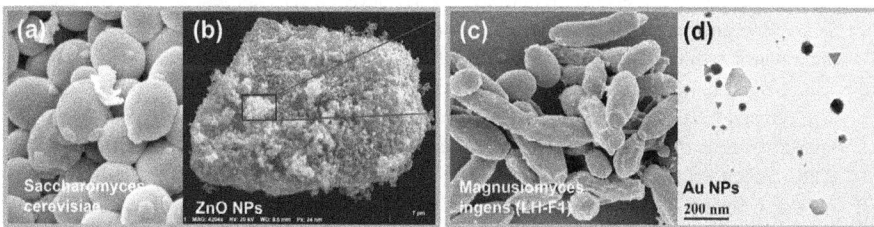

Figure 4.7: (a) Yeast *Saccharomyces cerevisiae*; **(b)** SEM image of ZnO nanoparticle (adapted from ref. [70]); **(c)** yeast *Magnusiomyces ingens* (LH-F1); and **(d)** TEM image of Au NPs (adapted from ref. [72]).

4.5 References

[1]   A. Rahman, M. H. Harunsani, A. L. Tan and M. M. Khan. Zinc oxide and zinc oxide-based nanostructures: Biogenic and phytogenic synthesis, properties and applications, *Bioprocess Biosyst. Eng.*, 2021, **44**, 1333–1372.

[2]   K. Shivaji, E. S. Monica, A. Devadoss, D. D. Kirubakaran, C. R. Dhas, S. M. Jain and S. Pitchaimuthu, Synthesizing green photocatalyst using plant leaf extract for water pollutant treatment: in *Green Photocatalysts*, eds. M. Naushad, S. Rajendran and E. Lichtfouse Springer International Publishing, Gewerbestrasse, 2020, pp. 25–46.

[3]   Y. W. Getahun, J. Gardea-Torresdey, F. S. Manciu, X. Li and A. A. El-Gendy. Green synthesized superparamagnetic iron oxide nanoparticles for water treatment with alternative recyclability, *J. Mol. Liq.*, 2022, **356**, 118983.

[4]   H. S. Kamaluddin and K. Narasimharao, Enviro-friendly nanomaterial synthesis and its utilization for water purification: in *Novel Materials and Water Purification: Towards a Sustainable Future*, in *Novel Materials and Water Purification: Towards a Sustainable Future*, eds. G. L. Kyriakopoulos and M. G. Zamparas, Royal Society of Chemistry, Croydon, 2024, pp. 298–352.

[5]   T. Varadavenkatesan, R. Vinayagam and R. Selvaraj. Green synthesis and structural characterization of silver nanoparticles synthesized using the pod extract of *Clitoria ternatea* and its application towards dye degradation, *Mater. Today Proc.*, 2020, **23**, 27–29.

[6]   K. Tran Khac, H. Hoang Phu, H. Tran Thi, V. Dinh Thuy and H. Do Thi. Biosynthesis of silver nanoparticles using tea leaf extract (*Camellia sinensis*) for photocatalyst and antibacterial effect, *Heliyon*, 2023, **9**, e20707.

[7]   A. I. Lukman, B. Gong, C. E. Marjo, U. Roessner and A. T. Harris. Facile synthesis, stabilization, and anti-bacterial performance of discrete Ag nanoparticles using *Medicago sativa* seed exudates, *J. Colloid Interface Sci.*, 2011, **353**, 433–444.

[8]   M. M. Riyanto, L. Asysyafiiyah, M. I. Sirajuddin and N. Cahyandaru Direct synthesis of lemongrass (*Cymbopogon citratus* L.) essential oil-silver nanoparticles (EO-AgNPs) as biopesticides and application for lichen inhibition on stones, *Heliyon*, 2022, **8**, e09701.

[9]   O. S. ElMitwalli, O. A. Barakat, R. M. Daoud, S. Akhtar and F. Z. Henari. Green synthesis of gold nanoparticles using cinnamon bark extract, characterization, and fluorescence activity in Au/eosin Y assemblies, *J. Nanoparticle Res.*, 2020, **22**, 309.

[10]   V. T. Le, N. N. Q. Ngu, T. P. Chau, T. D. Nguyen, V. T. Nguyen, T. L. H. Nguyen, X. T. Cao and V.-D. Doan. Silver and gold nanoparticles from *Limnophila rugosa* leaves: Biosynthesis, characterization, and catalytic activity in reduction of nitrophenols, *J. Nanomater.*, 2021, **2021**, 5571663.

[11]   T. H. A. Nguyen, T. T. V. Le, B. A. Huynh, N. V. Nguyen, V. T. Le, V.-D. Doan, V. A. Tran, A.-T. Nguyen, X. T. Cao and Y. Vasseghian. Novel biogenic gold nanoparticles stabilized on poly(styrene-*co*-maleic anhydride) as an effective material for reduction of nitrophenols and colorimetric detection of Pb(II), *Environ. Res.*, 2022, **212**, 113281.

[12]   A. Saka, Y. Shifera, L. T. Jule, B. Badassa, N. Nagaprasad, R. Shanmugam, L. Priyanka Dwarampudi, V. Seenivasan and K. Ramaswamy. Biosynthesis of TiO$_2$ nanoparticles by Caricaceae (Papaya) shell extracts for antifungal application, *Sci. Rep.*, 2022, **12**, 15960.

[13]   A. Khan, A. Raza, A. Hashem, G. Dolores Avila-Quezada, E. Fathi Abd_Allah, F. Ahmad and A. Ahmad. Green fabrication of titanium dioxide nanoparticles via *Syzygium cumini* leaves extract: Characterizations, photocatalytic activity and nematicidal evaluation, *Green Chem. Lett. Rev.*, 2024, **17**, 2331063.

[14]   S. Pavithra, T. C. Bessy, M. R. Bindhu, R. Venkatesan, R. Parimaladevi, M. M. Alam, J. Mayandi and M. Umadevi. Photocatalytic and photovoltaic applications of green synthesized titanium oxide (TiO2) nanoparticles by *Calotropis gigantea* extract, *J. Alloys Compd.*, 2023, **960**, 170638.

[15] A. A. Kashale, K. P. Gattu, K. Ghule, V. H. Ingole, S. Dhanayat, R. Sharma, J.-Y. Chang and A. V. Ghule. Biomediated green synthesis of TiO2 nanoparticles for lithium ion battery application, *Compos. Part B Eng.*, 2016, **99**, 297–304.

[16] K. Rambabu, G. Bharath, F. Banat and P. L. Show. Green synthesis of zinc oxide nanoparticles using *Phoenix dactylifera* waste as bioreductant for effective dye degradation and antibacterial performance in wastewater treatment, *J. Hazard. Mater.*, 2021, **402**, 123560.

[17] C. Vidya., C. Manjunatha, M. N. Chandraprabha, M. Rajshekar and A. Raj.M.A.L. Hazard free green synthesis of ZnO nano-photo-catalyst using *Artocarpus heterophyllus* leaf extract for the degradation of Congo red dye in water treatment applications, *J. Environ. Chem. Eng.*, 2017, **5**, 3172–3180.

[18] V. B. Raghavendra, S. Shankar, M. Govindappa, A. Pugazhendhi, M. Sharma and S. C. Nayaka. Green synthesis of zinc oxide nanoparticles (ZnO NPs) for effective degradation of dye, polyethylene and antibacterial performance in waste water treatment, *J. Inorg. Organomet. Polym. Mater.*, 2022, **32**, 614–630.

[19] M. S. H. Bhuiyan, M. Y. Miah, S. C. Paul, T. Das Aka, O. Saha, M. M. Rahaman, M. J. I. Sharif, O. Habiba and M. Ashaduzzaman. Green synthesis of iron oxide nanoparticle using *Carica papaya* leaf extract: Application for photocatalytic degradation of Remazol yellow RR dye and antibacterial activity, *Heliyon*, 2020, **6**, e04603.

[20] N. Madubuonu, S. O. Aisida, A. Ali, I. Ahmad, T. Zhao, S. Botha, M. Maaza and F. I. Ezema. Biosynthesis of iron oxide nanoparticles via a composite of *Psidium guavaja-Moringa oleifera* and their antibacterial and photocatalytic study, *J. Photochem. Photobiol. B Biol.*, 2019, **199**, 111601.

[21] I. Ashraf, N. B. Singh and A. Agarwal. Green synthesis of iron oxide nanoparticles using Amla seed for Methylene blue dye removal from water, *Mater. Today Proc.*, 2023, **72**, 311–316.

[22] I. M. Rashid, S. D. Salman, A. K. Mohammed and Y. S. Mahdi. Green synthesis of nickle oxide nanoparticles for adsorption of dyes, *Sains Malaysiana*, 2022, **2**, 533–546.

[23] T. Riaz, A. Munnwar, T. Shahzadi, M. Zaib, S. Shahid, M. Javed, S. Iqbal, K. Rizwan, M. Waqas, B. Khalid, N. S. Awwad, H. A. Ibrahium and M. A. Bajaber. Phyto-mediated synthesis of nickel oxide (NiO) nanoparticles using leaves' extract of *Syzygium cumini* for antioxidant and dyes removal studies from wastewater, *Inorg. Chem. Commun.*, 2022, **142**, 109656.

[24] C. Kalita and P. Saikia. Magnetically separable tea leaf mediated nickel oxide nanoparticles for excellent photocatalytic activity, *J. Indian Chem. Soc.*, 2021, **98**, 100213.

[25] A. Wibowo, G. U. N. Tajalla, M. A. Marsudi, G. Cooper, L. A. T. W. Asri, F. Liu, H. Ardy and P. J. D. S. Bartolo Green Synthesis of Silver Nanoparticles Using Extract of Cilembu Sweet Potatoes (Ipomoea batatas L var. Rancing) as Potential Filler for 3D Printed Electroactive and Anti-Infection Scaffolds, *Molecules*, 2021, **26**, 2042.

[26] A. Maghfirah, M. M. Ilmi, A. T. N. Fajar and G. T. M. Kadja. A review on the green synthesis of hierarchically porous zeolite, *Mater. Today Chem.*, 2020, **17**, 100348.

[27] G. T. M. Kadja, I. R. Kadir, A. T. N. Fajar, V. Suendo and R. R. Mukti. Revisiting the seed-assisted synthesis of zeolites without organic structure-directing agents: Insights from the CHA case, *RSC Adv.*, 2020, **10**, 5304–5315.

[28] P. Solanki, V. Gupta and R. Kulshrestha. Synthesis of zeolite from fly ash and removal of heavy metal ions from newly synthesized zeolite, *E-Journal Chem.*, 2010, **7**, 356150.

[29] N. Murayama, H. Yamamoto and J. Shibata. Mechanism of zeolite synthesis from coal fly ash by alkali hydrothermal reaction, *Int. J. Miner. Process.*, 2002, **64**, 1–17.

[30] J. L. X. Hong, T. Maneerung, S. N. Koh, S. Kawi and C.-H. Wang. Conversion of coal fly ash into zeolite materials: synthesis and characterizations, process design, and its cost-benefit analysis, *Ind. Eng. Chem. Res.*, 2017, **56**, 11565–11574.

[31] D. Längauer, V. Čablík, S. Hredzák, A. Zubrik, M. Matik and Z. Danková. Preparation of synthetic zeolites from coal fly ash by hydrothermal synthesis, *Mater.* 2021, **14**, 1267.

[32] X. Ren, R. Qu, S. Liu, H. Zhao, W. Wu, H. Song, C. Zheng, X. Wu and X. Gao. Synthesis of zeolites from coal fly ash for removal of harmful gaseous pollutants: a review, *Aerosol Air Qual. Res.*, 2020, **20**, 1127–1144.

[33] K. Ojha, N. C. Pradhan and A. N. Samanta. Zeolite from fly ash: Synthesis and characterization, *Bull. Mater. Sci.*, 2004, **27**, 555–564.

[34] X. Ren, L. Xiao, R. Qu, S. Liu, D. Ye, H. Song, W. Wu, C. Zheng, X. Wu and X. Gao. Synthesis and characterization of a single phase zeolite A using coal fly ash, *RSC Adv.*, 2018, **8**, 42200–42209.

[35] P. Zhang, S. Li, P. Guo and C. Zhang. Seed-assisted, OSDA-free, solvent-free synthesis of ZSM-5 zeolite from iron ore tailings, *Waste Biomass Valori.*, 2020, **11**, 4381–4391.

[36] R. M. Mohamed, I. A. Mkhalid and M. A. Barakat. Rice husk ash as a renewable source for the production of zeolite NaY and its characterization, *Arab. J. Chem.*, 2015, **8**, 48–53.

[37] C. G. Flores, H. Schneider, J. S. Dornelles, L. B. Gomes, N. R. Marcilio and P. J. Melo. Synthesis of potassium zeolite from rice husk ash as a silicon source, *Clean. Eng. Technol.*, 2021, **4**, 100201.

[38] C. D. D. Sundari, S. Setiadji, Y. Rohmatullah, S. Sanusi, D. F. Nurbaeti, I. Novianti, I. Farida, A. Nurohmah and A. L. Ivansyah. Synthesis of zeolite L using rice husk ash silica for adsorption of Methylene blue: Kinetic and adsorption isotherm. *MATEC Web Conf*, 2018, **197**, 05002.

[39] C. Zhang, S. Li and S. Bao. Sustainable synthesis of ZSM-5 zeolite from rice husk ash without addition of solvents, *Waste Biomass Valori.*, 2019, **10**, 2825–2835.

[40] J. Madhu, A. Santhanam, M. Natarajan and D. Velauthapillai. CO_2 adsorption performance of template free zeolite A and X synthesized from rice husk ash as silicon source, *RSC Adv.*, 2022, **12**, 23221–23239.

[41] D. Prasetyoko, Z. Ramli, S. Endud, H. Hamdan and B. Sulikowski. Conversion of rice husk ash to zeolite beta, *Waste Manag.*, 2006, **26**, 1173–1179.

[42] T. Sriatun and L. Suyati. Synthesis of zeolite from sugarcane bagasse ash using cetyltrimethylammonium bromide as structure directing agent, *Indones. J. Chem.*, 2018, **18**, 159–165.

[43] C. Belviso, M. Mancinelli and A. Lettino. A green process for zeolite synthesis: Low-temperature vapor phase treatment of natural bauxites, *J. Mater. Sci.*, 2022, **57**, 16619–16631.

[44] E.-P. Ng, J.-H. Chow, R. R. Mukti, O. Muraza, T. C. Ling and K.-L. Wong. Hydrothermal synthesis of zeolite A from bamboo leaf biomass and its catalytic activity in cyanoethylation of methanol under autogenic pressure and air conditions, *Mater. Chem. Phys.*, 2017, **201**, 78–85.

[45] L. Ayele, E. Pérez, Á. Mayoral, Y. Chebude and I. Díaz. Synthesis of zeolite A using raw kaolin from Ethiopia and its application in removal of Cr(III) from tannery wastewater, *J. Chem. Technol. Biotechnol.*, 2018, **93**, 146–154.

[46] L. Ayele, J. Pérez-Pariente, Y. Chebude and I. Díaz. Conventional versus alkali fusion synthesis of zeolite A from low grade kaolin, *Appl. Clay Sci.*, 2016, **132–133**, 485–490.

[47] N. Sazali, Z. Harun, T. Abdullahi, N. H. Kamarudin, N. Sazali, M. R. Jamalludin, S. K. Hubadillah and S. S. Alias. The route of hydrothermal synthesis zeolite-A from the low-grade perak kaolin, Malaysia, *Silicon*, 2022, **14**, 7257–7273.

[48] P. M. Pereira, B. F. Ferreira, N. P. Oliveira, E. J. Nassar, K. J. Ciuffi, M. A. Vicente, R. Trujillano, V. Rives, A. Gil, S. Korili and E. H. De Faria, Synthesis of Zeolite A from Metakaolin and Its Application in the Adsorption of Cationic Dyes, *Appl. Sci.*, 2018, **8**, 608.

[49] T. S. Jamil, H. S. Ibrahim, I. H. Abd El-Maksoud and S. T. El-Wakeel. Application of zeolite prepared from Egyptian kaolin for removal of heavy metals: I. Optimum conditions, *Desalination*, 2010, **258**, 34–40.

[50] T. S. Jamil, H. H. Abdel Ghafar, H. S. Ibrahim and I. H. Abd El-Maksoud, removal of Methylene blue by two zeolites prepared from naturally occurring Egyptian kaolin as cost effective technique, *Solid State Sci.*, 2011, **13**, 1844–1851.

[51] M. M. Selim and I. H. Abd El-Maksoud, Hydrogenation of edible oil over zeolite prepared from local kaolin, *Microporous Mesoporous Mater.*, 2004, **74**, 79–85.

[52] Y. He, S. Tang, S. Yin and S. Li. Research progress on green synthesis of various high-purity zeolites from natural material-kaolin, *J. Clean. Prod.*, 2021, **306**, 127248.

[53] D. P. Hartati, M. Santoso, I. Qoniah, W. L. Leaw, P. B. D. Firda and H. Nur. A review on synthesis of kaolin-based zeolite and the effect of impurities, *J. Chinese Chem. Soc.*, 2020, **67**, 911–936.

[54] P. F. Corregidor, D. E. Acosta and H. A. Destéfanis. Kinetic study of seed-assisted crystallization of ZSM-5 zeolite in an osda-free system using a natural aluminosilicate as starting source, *Ind. Eng. Chem. Res.*, 2018, **57**, 13713–13720.

[55] X. Jiang, X. Fan, W. Xu, R. Zhang and G. Wu. Biosynthesis of bimetallic Au–Ag nanoparticles using *Escherichia coli* and its biomedical applications, *ACS Biomater. Sci. Eng.*, 2020, **6**, 680–689.

[56] K. Mehri, A. Parinaz, B. A. Leila and A.-D. Maryam. Biosynthesis of copper oxide nanoparticles using *Lactobacillus casei* subsp. casei and its anticancer and antibacterial activities, *Curr. Nanosci.*, 2020, **16**, 101–111.

[57] E. Coli. Symptoms, transmission, treatment & prevention, https://www.tuasaude.com/en/e-coli/, (accessed 24 August 2024).

[58] M. Rai, S. Bonde, P. Golinska, J. Trzcińska-Wencel, A. Gade, K. A. Abd-Elsalam, S. Shende, S. Gaikwad and A. P. Ingle, Fusarium as a Novel Fungus for the Synthesis of Nanoparticles: Mechanism and Applications, *J. Fungi.*, 2021, **7**, 139.

[59] A. Nyabadza, É. McCarthy, M. Makhesana, S. Heidarinassab, A. Plouze, M. Vazquez and D. Brabazon. A review of physical, chemical and biological synthesis methods of bimetallic nanoparticles and applications in sensing, water treatment, biomedicine, catalysis and hydrogen storage, *Adv. Colloid Interface Sci.*, 2023, **321**, 103010.

[60] K. Sneha, M. Sathishkumar, J. Mao, I. S. Kwak and Y.-S. Yun. *Corynebacterium glutamicum*-mediated crystallization of silver ions through sorption and reduction processes, *Chem. Eng. J.*, 2010, **162**, 989–996.

[61] Y. Qu, X. Li, S. Lian, C. Dai, Z. Jv, B. Zhao and H. Zhou. Biosynthesis of gold nanoparticles using fungus *Trichoderma* sp. WL-Go and their catalysis in degradation of aromatic pollutants, *IET Nanobiotechnology*, 2019, **13**, 12–17.

[62] E. I. Murillo-Rábago, A. R. Vilchis-Nestor, K. Juarez-Moreno, L. E. Garcia-Marin, K. Quester and E. Castro-Longoria optimized synthesis of small and stable silver nanoparticles using intracellular and extracellular components of fungi: An alternative for bacterial inhibition, *Antibiotics*, 2022, **11**, 800.

[63] F. Mirkhani and H. Alaei. Species diversity of indigenous trichoderma from alkaline pistachio soils in Iran, *Mycol. Iran.*, 2015, **2**, 22–37.

[64] K. Chehri, B. Salleh, T. Yli-Mattila, K. R. N. Reddy and S. Abbasi. Molecular characterization of pathogenic Fusarium species in cucurbit plants from Kermanshah province, Iran, *Saudi J. Biol. Sci.*, 2011, **18**, 341–351.

[65] N. Naimi-Shamel, P. Pourali and S. Dolatabadi. Green synthesis of gold nanoparticles using *Fusarium oxysporum* and antibacterial activity of its tetracycline conjugant, *J. Mycol. Med.*, 2019, **29**, 7–13.

[66] A. Ahmad, P. Mukherjee, S. Senapati, D. Mandal, M. I. Khan, R. Kumar and M. Sastry. Extracellular biosynthesis of silver nanoparticles using the fungus *Fusarium oxysporum*, *Colloids Surf. B: Biointerfaces*, 2003, **28**, 313–318.

[67] S. M. Husseiny, T. A. Salah and H. A. Anter. Biosynthesis of size controlled silver nanoparticles by *Fusarium oxysporum*, their antibacterial and antitumor activities, *Beni-Suef Univ. J. Basic Appl. Sci.*, 2015, **4**, 225–231.

[68] T. Khalafi, F. Buazar and K. Ghanemi. Phycosynthesis and enhanced photocatalytic activity of zinc oxide nanoparticles toward organosulfur pollutants, *Sci. Rep.*, 2019, **9**, 6866.

[69] G. Caliskan, T. Mutaf, H. C. Agba and M. Elibol. Green synthesis and characterization of titanium nanoparticles using microalga, *Phaeodactylum tricornutum*, *Geomicrobiol. J.*, 2022, **39**, 83–96.

[70] G. T. Tran, N. T. H. Nguyen, N. T. T. Nguyen, T. T. T. Nguyen, D. T. C. Nguyen and T. Van Tran. Formation, properties and applications of microalgae-based ZnO nanoparticles: A review, *J. Environ. Chem. Eng.*, 2023, **11**, 110939.

[71] R. Hachicha, F. Elleuch, H. Ben Hlima, P. Dubessay, H. de Baynast, C. Delattre, G. Pierre, R. Hachicha, S. Abdelkafi, P. Michaud and I. Fendri, Biomolecules from Microalgae and Cyanobacteria: Applications and Market Survey, *Appl. Sci.*, 2022, **12**, 1924.

[72] S. Skalickova, M. Baroň and J. Sochor. Nanoparticles biosynthesized by yeast: A review of their application, *Kvas. Prum.*, 2017, **63**, 290–292.

Chapter 5
Grand challenges and opportunities in scale-up of material synthesis

5.1 Introduction

Nanomaterials are present in a wide range of items, from drug delivery systems to advanced electronic devices. The unique physicochemical properties resulting from their small size and high specific surface area make a wide variety of nanomaterials crucial for research in academia and industry [1]. Despite their potential benefits, scaling up the production of nanomaterials for mass manufacturing has posed several scientific and technological challenges in the past [2]. This chapter discusses some of the major obstacles and opportunities that remain to be addressed to achieve successful large-scale synthesis of nanomaterials.

5.2 Utilization of nanomaterials in different industries

The utilization of nanomaterials holds vast potential across diverse fields and industrial domains, including information technology, national security, healthcare, telecommunications, energy production, food security, and environmental protection [3]. Nanomaterials facilitate the production of reusable, long-lasting "smart fabrics" embedded with elastic nanosensors and processors capable of overseeing healthcare monitoring, optimizing solar energy utilization, and generating power. Furthermore, the increasing use of various nanomaterials in catalysis is not just about enhancing chemical reaction rates. It's about reducing the required catalytic materials, cutting costs, and most importantly, limiting pollution [4]. Nanotechnology is not just a buzzword, it's a game-changer in the medical field. It's expanding the range of medical instruments, information, and medicines accessible to practitioners. Nanomedicine, the use of nanomaterials to provide efficient results for preventive care, diagnostics, and treatment, is a promising frontier [5].

5.3 Common nanomaterials used in different industrial sectors

Due to their economic importance, metal oxides, especially silicon dioxide, titanium dioxide, aluminum oxide, and iron oxide, are among the most extensively utilized inorganic nanomaterials in various industrial sectors. Metal oxide nanomaterials find application in sectors such as electronics, pharmaceuticals, skin care, biochemistry, and catalysis. Bottom-up techniques are frequently employed to synthesize nanosized metal oxides [6].

https://doi.org/10.1515/9783111316819-005

Eliminating transition metals in dilute solutions is a prevalent mass-production technique for these nanomaterials. Additionally, quantum dots are another highly utilized nanomaterial in the industry, particularly electronics. These are tiny particles of matter with attributes that change when adding a single electron. Quantum dots are composed of a few hundred atoms of semiconducting material, which, when excited, emit light at varying wavelengths based on their size, rendering them particularly valuable indicators of cellular activity in biomedicine [7]. Metallic nanoparticles, like gold and silver, are also widely applied in the healthcare industry [8]. Normally, foreign object penetration results in cell damage or death, but metallic nanoparticles can pass through cellular membranes without causing harm. Consequently, they serve as excellent drug carriers to either normal tissues or radiation in malignant cells. A prevalent synthetic technique for producing metallic nanoparticles involves reducing metal salts in a suspension with the aid of a modifier. A wide array of reductants and modifiers have been discovered, enabling the production of substantial quantities of nanoparticles. Recently, the idea of nanofluids has gained acceptance within the industrial world, leading to a race to develop technologies for commercializing nanofluids [9].

5.4 Nanomaterials synthesis on a laboratory scale

Conventional nanomaterial synthesis methods employed in laboratories are often labor-intensive, time-consuming, and yield relatively low quantities. While these small amounts can suffice for initial research, they hinder the transition of promising innovations into practical applications. Industrial nanomaterial production typically involves two primary approaches: bottom-up techniques, which construct molecule-by-molecule, and top-down techniques, which reduce larger raw materials to nanoscale dimensions [10, 11]. Top-down methods entail fragmentation (splitting up), tearing, and cleaving larger raw materials into nanomaterial form. Advanced lithography and etching techniques can be used not only to remove material but also to shape nanomaterial into specific surface structures. Some examples include stereolithography, nanopatterning, electron beam lithography, ion beam etching, and subsurface reactive ion etching. Bottom-up approaches involve constructing the atomic structure from scratch by placing atoms under certain conditions to form the desired configuration. Common bottom-up processes include chemical vapor deposition, physical vapor deposition, and atomic layer deposition. Other chemical synthesis methods encompass hydrothermal, sol-gel, chemical reduction, plasma-assisted production, condensation, electrochemical deposition methods, and aerosol techniques [12].

5.5 Scaling up nanomaterials: importance and challenges

The process of successfully scaling up from laboratory-scale research to production begins with assembling the right team. A team requiring a diverse range of skills is necessary, including experts in chemistry, chemical engineering, analytical chemistry, and environmental health and safety (EHS). The team must formulate a comprehensive plan for the entire scaling-up process. As products are introduced into the industrial environment, the queries requiring answers have diversified. Raw materials with identical specifications can be obtained from different sources; however, they may perform differently when specifications vary. The role of a chemist initially involves understanding basic chemistry and collaborating with the product development or application team to determine how a product should perform in a specific application. The chemist requires hands-on experience in the lab with the reaction to identify the causes of plant upsets and most importantly, know how to address them effectively. This involves producing materials near or beyond the product specifications. During these runs, the chemist needs to understand what factors influence the product and assess the robustness of the synthesis. For instance, if a reaction requires a temperature of 150 °C, it's important to know what happens if the plant's control system maintains a range of 145–155 °C. Will the product have the desired conversion? When running these trial lab batches, the chemist should consider all the product specifications. To ensure a successful scale-up, a product should have clearly defined analytical targets and appropriate ranges for those targets. As the product is scaled up to the plant, it needs to be analyzed using equipment available at the plant and the methods that the plant utilizes. The specifications set by the team must fall within the accuracy limits of the analytical tests being used. Before scale-up, the team needs to ensure that the analytical methods exist in the production environment. When developing batch sizes, the chemist must consider which materials are most challenging to handle in the process. Large-scale production of nanomaterials is needed to replace existing bulk materials with better-capable nanomaterials. Nanoscience offers a unique opportunity to understand physio-chemical phenomena at the most fundamental level, which, in theory, should be useful in improving and optimizing a structure under investigation. Despite these basic achievements in the field, nanotechnology is still facing significant challenges [13]. This is because translating scientific findings published in research journals into industrial technology applications is still a cumbersome task. The issue is multifaceted. First, the properties of materials alter significantly when scaling up or down, especially when moving between the nanoscale, mesoscale, and macroscale. Second, the industry is reluctant to invest heavily in developing new large-scale techniques for manufacturing nanomaterials unless a substantial profit is guaranteed. This dilemma is particularly crucial in applied sciences due to the gap between laboratory research and industrial-scale research.

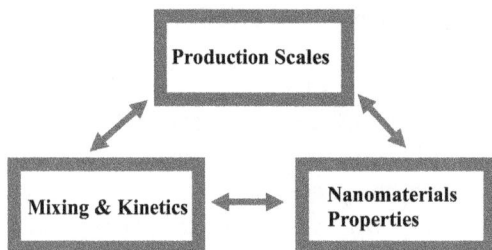

Figure 5.1: Schematic representation of interconnectivity of knowledge of the process chemistry, scale-up, and nanomaterials properties.

Several previous research investigations have reported that additive chemistry, precursor chemistry, and chemical conditions strongly influence the physicochemical properties of nanomaterials [14]. However, the knowledge required to design nanomaterials, such as the relationship between synthesis-structure-property, as well as a thorough understanding of the kinetics of the process for these materials, is not readily available. Future research needs to address the lack of precise criteria necessary to design commercially relevant nanomaterials. To significantly impact the production and commercial use of nanomaterials, further research is required for precise control over their properties, scalable production methods, and advanced processing techniques (as shown in Figure 5.1) [15]. Undoubtedly, nanomaterials outperform conventional bulk materials and are highly sought-after [16]. All the methods described in the previous chapters can be classified into two categories: bottom-up and top-down processes. Bottom-up methods start from a dissolved molecule and proceed to a precipitate, whereas top-down methods start with a macro-sized drug powder and reduce it into smaller particles. Different industries widely adopt the latter approach, while the former approach is less popular at an industrial level due to the difficulty of removing the traces of any remaining solvent involved in the process [17]. Several factors need to be considered during the scale-up of nanomaterials, from the laboratory to the market. For example, one must consider the nature of the material, its Generally Regarded as Safe (GRAS) status, the toxicological features associated with the size and shape of nanoparticles [18], the biodegradability of nanomaterials, and the balance of a multi-component system at a large scale. Careful selection of materials, solvents, nanomaterial development procedures, cost, and the acceptability of the final product are crucial factors to consider. During the scale-up of laboratory methods, there may be instances where the desired characteristics of nanomaterials are lost. For example, in a study on the scale-up of nanoparticles prepared using an emulsion method, it was observed that an increase in impeller speed and agitation time reduced the particle size, while the entrapment efficiency remained unchanged [19]. From a scale-up perspective, the choice of a nanomaterial production method is important for saving time, especially during pilot batch production. In a comparative study of ibuprofen-loaded nanoparticles, the nanoprecipitation method was found to require significantly less time (approximately 2 h) compared to the emulsion-based method (about 3 h) for nanoparticle production [20].

5.6 Factors that influence a scale-up

Transitioning a product from benchtop scale to large-scale production is not merely multiplying reactant quantities by 200-times. Most laboratory experiments are designed to facilitate rapid heat dissipation from the reaction vessel. On the other hand, heat-up and cool-down cycles are considerably longer in production-scale vessels. Additionally, extended reaction times may affect a product's properties. If the reaction is performed in glassware, excess vapor may be vented to the atmosphere or flow through a condenser, mitigating any pressure concerns. However, if laboratory work is conducted in pressure vessels, the vessels are commonly rated to withstand pressures well above any anticipated levels. Therefore, it is crucial to investigate and document the vapor pressure of the reaction mixture throughout the entire process. Changes in temperature can significantly impact the viscosity of the material, which can be difficult to measure during the entire process. However, the chemist can measure the starting materials' viscosity and the final product's viscosity as a comparison point [21]. Chemical engineers need to understand the viscosity profile of a reaction over its course. For instance, a raw material with water viscosity may function effectively at room temperature in a reactor's pump. However, as the temperature escalates, the viscosity may become insufficient to adequately lubricate the pump seals in the reactor [22]. If the product's viscosity increases over the course of reaction, it may flow smoothly through the pump at elevated temperatures but might become too viscous for the pump to circulate the material when temperature decreases properly.

5.7 Safety considerations

In a laboratory setting, when a product is being developed, all the work is usually carried out in a fume hood. However, in a production setting, the ventilation is typically natural ventilation, taking advantage of the wind and air movement around the vessels. As the process moves towards scale-up, the EHS group will examine both the product and the processes involved. It is crucial for the EHS group to identify if the product and any isolated intermediate materials are considered toxic or hazardous in nature. In cases where these materials have not yet been produced at an industrial scale, it is recommended to file a preliminary manufacturing notification with the concerned authorities [23]. Another important consideration is compliance with the air emission permits of the production facilities. Common chemicals regularly employed in a laboratory setting may surpass the air emission limits if they are released into the atmosphere during chemical reactions. Even simple heating of water can potentially exceed the pressure rating of production vessels if water is present in high concentrations. Lower boiling point solvents can further aggravate this effect [24]. When conducting the Process Safety Analysis for a new product, it is crucial to consider several safety-related factors related to the production process. These topics can include discussing the maximum allowable working pressure and vessel pressure ratings. A member of the EHS team, which stands for Health, Safety, and Environment, will assess material compatibility, vessel pressure and temperature ratings, control systems, plant safety interlocks, relief device settings, and other potential risk factors. Typically, the technical team responsible for introducing the product will participate as well. The primary objective is to ensure that the new production process aligns with the capabilities of the existing plant equipment in terms of safety. In cases where the product or process falls outside these capabilities, recommendations are made to enhance the equipment. Protocols concerning these changes may differ from one production facility to another; however, all of them undergo some type of management of change process when new materials are introduced into the plant. Successful completion of the initial production run can be achieved through various means. Achieving success in a production environment can manifest in different ways, ranging from completing the first large-scale production run of a product to successfully producing millions of pounds of the product for a customer. Regardless of the chosen definition, this achievement can only be realized through a team effort, which involves the cooperation of multiple departments, including chemists, chemical engineers, EHS specialists, supply chain personnel, and more. The contributions of chemists in kickstarting the project, by defining targets and verifying the process, are crucial [25]. Once the process is underway, the collaboration of other departments can provide valuable assistance in evaluating and prioritizing all the factors and safety considerations necessary to ensure a successful scale-up production run.

5.8 Leading technologies for scaling up nanomaterials

For instance, Cerion [26] implemented a Design for Manufacturing technique to facilitate the scale-up process. This approach uses a phased-gate system to simplify, optimize, and enhance the nanomaterial while avoiding issues that may hinder scale-up or increase production costs. The technique is comprehensive, taking into account critical commercial, scientific, technical, environmental, supply chain, and logistical considerations that ultimately impact the final cost of the nanomaterial. Another advancement in the field comes from researchers at the University of South California, who have devised an automated production technique that can accelerate the time-consuming batch-based approach currently in use [27]. This method utilizes 3D-manufactured tubes with a diameter of 250 μm. The scientists evaluated this approach by passing non-mixing substances like oil and water through four parallel tubes simultaneously. The resistance encountered as the fluids passed through the pipes resulted in the formation of nanoparticle droplets at the tube outlets.

5.9 Synthesis of nanofluids and its industrial significance

The industrial-scale production of nanofluids can follow different routes, but they are all based on the commonly used laboratory-scale methods. Many industries producing nanofluids have progressed from their primary activity of manufacturing nanoparticles. Conversely, some companies have initiated the introduction of nanoparticles as additives to pre-existing formulations of fluids for specific applications, adding value to their products [28]. To produce nanofluids using this approach, manufacturers typically purchase expensive nanoparticles, which increases the overall cost of the synthesized nanofluid. While the one-step method provides convenience and advantageous ther-mos-physical properties for the final product, it may not be economically viable in industrial settings. The two-step preparation method of nanofluids poses the biggest challenge in terms of stabilization, requiring significant skills and understanding of the behavior of nanofluids [29]. Ultrasonication leverages cavitation phenomena to achieve a uniform and stable dispersion of nanoparticles in liquids. Many scientists have utilized ultrasonication technique for preparing nanofluids in laboratory-scale studies [30]. While ultrasonication has proven to be very effective in these studies, its application in industrial-scale production is limited. An alternative approach to generating the same cavitation effects is through the use of hydrodynamic cavitation processes, which are more feasible for industrial-scale production. Radkar et al. [31] successfully synthesized water-based ZnO nanofluids using hydrodynamic cavitation. The resulting nanofluids exhibited superior thermal properties. Similarly, Kiu et al. [32] prepared nano lubricants by dispersing carbon nanotubes using hydrodynamic cavitation. They found better lubricating characteristics in the nanolubricant prepared via hydrodynamic cavitation compared to that prepared using ultrasonic cavitation. The potential for scalability in the process of hydrodynamic cavitation is more favorable than that for ultrasonic cavitation. While methods like ball milling, mechanical stirring, and homogenization can be considered for large-scale production of nanofluids, each has its own set of drawbacks. Ball milling, although easy and widely used in various industrial applications, consumes a significant amount of energy and is considered inefficient for nanofluid preparation [33]. Mechanical stirring, on the other hand, can be achieved through magnetic stirring, it has several drawbacks and has been found to be quite ineffective in completely breaking them down. Even if mechanical stirring is successful in producing a nanofluid with minimal agglomerates, the process can take several hours and consume a significant amount of energy. On the other hand, a high-pressure homogenizer can be a more suitable option for the stable preparation of nanofluids, offering advantages over mechanical stirring in terms of efficiency and resource utilization. The popularity of large-scale homogenizers in industries is primarily due to their availability and widespread use for various applications. Additionally, the use of stabilizers and dispersants can further enhance the stability of nanofluids [34]. It is crucial to evaluate the feasibility of each step for production scalability. Establishing certified guidelines for large-scale productions is essential and could be very helpful. The development of such techniques and skills is still ongoing [35]. However, there are challenges in using nano-fluids at a large scale, such as higher pumping power costs, uncertain stability issues, and potential detrimental effects on pipeline and equipment materials due to corrosion and erosion. While their versatility makes them highly sought after, they tend to have a high price tag. The nanoparticles of such materials are inherently expensive, adding to the overall cost. However, once issues related to scalability, stability, and compatibility with materials are addressed, there is still the significant concern of nanoparticles' tox-icity to consider [23]. With the increased need for large-scale applications, managing the disposal of significant amounts of nanofluids becomes a critical issue.

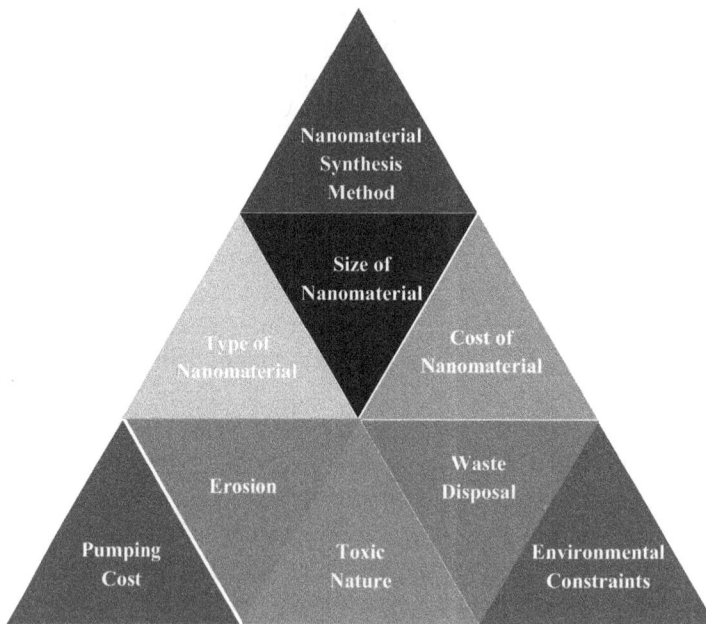

Figure 5.2: Challenges in scale-up of nanofluid synthesis.

Countries with stringent regulations regarding environmental protection often establish strict standards to prevent the production and entry of specific materials into their terri-tory. This can present significant challenges in the production of nanomaterials and in their subsequent application. Considering all the issues mentioned previously, a coun-try's research and development strategies ultimately impact the feasibility of scaling up and commercializing these technologies. Figure 5.2 illustrates the challenges related to scaling up nanofluids. Based on multiple studies, increasing the concentration of a nano-fluid can enhance its thermal performance, which is influenced by various mechanisms [36]. However, increased concentration directly corresponds to an increase in the cost of the nanofluid due to the greater amount of nanoparticles required. Hemmat Esfe et al.'s [37] study analyzed the cost of a single-walled carbon nanotube-zinc oxide (SWCMT-ZnO) nanofluid dispersed in an ethylene glycol-water mixture, and they justified the economic feasibility of this specific nanofluid type by comparing it to nanofluids containing individual nanoparticles of SWCNT and ZnO. While ZnO nanoparticles are less expensive and more affordable than SWCNT, the addition of SWCNT to a ZnO nanofluid results in improved thermal performance at a lower cost compared to SWCNT nanofluids. The same research group has conducted similar comparative studies on various other types of nanofluids [38].

5.10 Scale-up of nanofluids synthesis

An essential aspect of any synthesis process is its scalability, ensuring that the desirable physicochemical properties of the best catalyst identified through laboratory tests remain consistent during scaling up. While numerous synthesis methods encounter considerable obstacles, flame aerosol synthesis has already been established as a scalable technology, as exemplified by the industrial-scale production of photocatalyst TiO_2 [39]. The process parameters used strongly affect a material's physicochemical properties. Therefore, a thorough understanding of the process, based on both experimental observations and theoretical simulations, is crucial for successfully scaling up the process. Grohn et al. [40] demonstrated recently that ZrO_2 nanoparticles production could be scaled up from 100 to 500 g h^{-1}, without compromising important product properties such as crystal size and type. This success was achieved by maintaining a constant high-temperature residence time during the synthesis. The feasibility of producing materials at large scales, such as SiO_2 (41,000 g h^{-1}), ZrO_2 (600 g h^{-1}), and Y_2O_3/ZrO_2 (4,300 g h^{-1}), has already been demonstrated by various authors, as detailed in a recent study [40]. Furthermore, the production of ZnO nanorods can also be achieved at high production rates, specifically 43 kg h^{-1} [41]. It appears reasonable to suggest that the large-scale production of multi-component catalysts, maintaining their physicochemical properties, is feasible using flame aerosol synthesis. This method can potentially replace time-consuming wet-synthesis techniques in many cases. Pilot-scale production of a binary catalyst (V_2O_5/TiO_2) has already been demonstrated using a diffusion flame reactor, with a high production rate of 200 g h^{-1} [42]. Notably, the catalyst's performance in removing NO was found to be superior compared to the catalyst with a similar composition derived through classical wet-chemistry methods.

5.11 Concluding remarks

Nanomaterials synthesis is a commonly utilized process in various industrial sectors and is being actively researched in academic and industrial laboratories for a wide range of applications, including catalysis. The development of a scalable and cost-effective process for nanomaterial synthesis poses significant challenges for process chemists, particularly during the initial scale-up of the synthesis process. Scaling up the production of materials can be a significant challenge for many synthesis techniques, particularly when there is limited understanding of the complex dynamics involved in producing homogeneous multi-component materials. Scaling up can, however, be a highly efficient and cost-effective approach for mass-producing nanomaterials for various industrial applications. To ensure that cost-effective materials can be consistently and successfully produced, it is crucial to follow a carefully developed procedure when transferring nanomaterial technologies from laboratory to industrial settings. Despite ongoing developments in large-scale production techniques, there remain challenges to overcome. One such challenge is the high cost of ideal precursors, such as alkoxides and organometallic compounds. While cheaper nitrate precursors are preferred, they tend to produce inhomogeneous particles, which are not always desired. However, through the use of appropriate chemical methods, it is possible to produce homogeneous catalytic particles using these cheaper precursors, opening up new opportunities for the production of materials at lower costs. Establishing a standard methodology for the initial characterization of nanomaterials is crucial as it impacts the analysis of particle growth processes, affecting their performance. Modifying the surface of nanoparticles with functional groups during the production process is a further aspect that can be explored for potential cost reduction. The successful implementation of such processes can significantly reduce the overall cost of the production process. Currently, only a limited number of studies demonstrate the feasibility of scaling up nanomaterial synthesis using optimized techniques. While some progress has been made in understanding the chemical and physical processes involved in synthesis methodologies for advanced materials, there is still more work to be done to harness the potential of various methodologies in this field fully. A collaborative effort among material scientists, chemists, physicists, and engineers is necessary to further research in this area and advance the synthesis of more effective and efficient materials.

5.12 References

[1]   M. M. El-Kady, I. Ansari, C. Arora, N. Rai, S. Soni, D. K. Verma, P. Singh and A. E. D. Mahmoud. Nanomaterials: A comprehensive review of applications, toxicity, impact, and fate to environment, *J. Mol. Liq.*, 2023, **370**, 121046.

[2]   I. Khan, K. Saeed and I. Khan. Nanoparticles: Properties, applications and toxicities, *Arab. J. Chem.*, 2019, **12**, 908–931.

[3]   N. Joudeh and D. Linke. Nanoparticle classification, physicochemical properties, characterization, and applications: A comprehensive review for biologists, *J. Nanobiotechnology.*, 2022, **20**, 262.

[4]   M. A. Gatoo, S. Naseem, M. Y. Arfat, A. Mahmood Dar, K. Qasim and S. Zubair. Physicochemical properties of nanomaterials: Implication in associated toxic manifestations, *Biomed Res. Int.*, 2014, **2014**, 498420.

[5]   V. Gubala, L. J. Johnston, Z. Liu, H. Krug, C. J. Moore, C. K. Ober, M. Schwenk and M. Vert. Engineered nanomaterials and human health: Part 1. preparation, functionalization and characterization (IUPAC Technical Report), *Pure and Applied Chemistry*, 2018, **90**, 1283–1324.

[6]   M. Rahman, K. S. Islam, T. M. Dip, M. F. M. Chowdhury, S. R. Debnath, S. M. M. Hasan, M. S. Sakib, T. Saha, R. Padhye and S. Houshyar. A review on nanomaterial-based additive manufacturing: Dynamics in properties, prospects, and challenges, *Prog. Addit. Manuf.*, 2024, **9**, 1197–1224.

[7]   V. G. Reshma and P. V. Mohanan. Quantum dots: Applications and safety consequences, *J. Lumin.*, 2019, **205**, 287–298.

[8]   A. Meher, A. Tandi, S. Moharana, S. Chakroborty, S. S. Mohapatra, A. Mondal, S. Dey and P. Chandra. Silver nanoparticle for biomedical applications: A review, *Hybrid Adv.*, 2024, **6**, 100184.

[9]   S. P. Tembhare, D. P. Barai and B. A. Bhanvase. Performance evaluation of nanofluids in solar thermal and solar photovoltaic systems: A comprehensive review, *Renew. Sustain. Energy Rev.*, 2022, **153**, 111738.

[10] T. M. Joseph, D. Kar Mahapatra, A. Esmaeili, Ł. Piszczyk, M. S. Hasanin, M. Kattali, J. Haponiuk and S. Thomas Nanoparticles: Taking a Unique Position in Medicine, *Nanomaterials.*, 2023, **13**, 574.

[11] N. Abid, A. M. Khan, S. Shujait, K. Chaudhary, M. Ikram, M. Imran, J. Haider, M. Khan, Q. Khan and M. Maqbool. Synthesis of nanomaterials using various top-down and bottom-up approaches, influencing factors, advantages, and disadvantages: A review, *Adv. Colloid Interface Sci.*, 2022, **300**, 102597.

[12] N. Baig, I. Kammakakam and W. Falath. Nanomaterials: A review of synthesis methods, properties, recent progress, and challenges, *Mater. Adv.*, 2021, **2**, 1821–1871.

[13] B. Elzein. Nano revolution: "Tiny tech, big impact: How nanotechnology is driving SDGs progress, *Heliyon.*, 2024, **10**, e31393.

[14] S. V. Patwardhan. Biomimetic and bioinspired silica: Recent developments and applications, *Chem. Commun.*, 2011, **47**, 7567–7582.

[15] M. C. Roco, C. A. Mirkin and M. C. Hersam. Nanotechnology research directions for societal needs in 2020: Summary of international study, *J. Nanoparticle Res.*, 2011, **13**, 897–919.

[16] V. Wagner, A. Dullaart, A.-K. Bock and A. Zweck. The emerging nanomedicine landscape, *Nat. Biotechnol.*, 2006, **24**, 1211–1217.

[17] J. Junghanns and R. Müller. Nanocrystal technology, drug delivery and clinical applications, *Int J Nanomedicine.*, 2008, **2**, 295–310.

[18] C. Buzea, I. Pacheco and K. Robbie. Nanomaterials and nanoparticles: Sources and toxicity, *Biointerphases, Biointerphases.*, 2007, **2**, MR17–71.

[19] A. Colombo, S. Briançon, J. Lieto and H. Fessi. Project, design, and use of a pilot plant for nanocapsule production, *Drug Dev Ind Pharm.*, 2001, **27**, 1063–1072.

[20] S. A. Galindo-Rodríguez, F. Puel, S. Briançon, E. Allémann, E. Doelker and H. Fessi. Comparative scale-up of three methods for producing ibuprofen-loaded nanoparticles, *Eur. J. Pharm. Sci.*, 2005, **25**, 357–367.

[21] A. Akpek, Detailed analysis of the effects of viscosity measurement errors caused by heat transfer during continuous viscosity measurements under various temperature changes and the proposed solution of a non-dimensional parameter called the Akpek Number, *Appl. Sci.*, 2023, **13**, 10684.

[22] Y. Park, G. Hong, S. Jun, J. Choi, T. Kim, M. Kang and G. Jang, Thermo-Fluid-Structural Coupled Analysis of a Mechanical Seal in Extended Loss of AC Power of a Reactor Coolant Pump, *Lubricants.*, 2024, **12**, 212.

[23] S. A. Mazari, E. Ali, R. Abro, F. S. A. Khan, I. Ahmed, M. Ahmed, S. Nizamuddin, T. H. Siddiqui, N. Hossain, N. M. Mubarak and A. Shah. Nanomaterials: Applications, waste-handling, environmental toxicities, and future challenges – A review, *J. Environ. Chem. Eng.*, 2021, **9**, 105028.

[24] M. J. Mitchell, M. M. Billingsley, R. M. Haley, M. E. Wechsler, N. A. Peppas and R. Langer. Engineering precision nanoparticles for drug delivery, *Nat. Rev. Drug Discov.*, 2021, **20**, 101–124.

[25] H.-D. Volk, M. M. Stevens, D. J. Mooney, D. W. Grainger and G. N. Duda. Key elements for nourishing the translational research environment, *Sci. Transl. Med.*, 2015, **7**, 282cm2–282cm2.
[26] C. A. Charitidis, P. Georgiou, M. A. Koklioti, A.-F. Trompeta and V. Markakis. Manufacturing nanomaterials: From research to industry, *Manuf. Rev.*, 2014, **1**, 11.
[27] R. Perkins. Here's a way to produce nanomaterials on a larger scale. Retrieved from USC News, https://news.usc.edu/92312/heres-a-way-to-producenanomaterials-on-a-larger-scale/Trofimencoff, T., (accessed 23 March 2024).
[28] Deepika. Nanotechnology implications for high performance lubricants, *SN Appl. Sci.*, 2020, **2**, 1128.
[29] C. Zamora-Ledezma, C. Narváez-Muñoz, V. H. Guerrero, E. Medina and L. Meseguer-Olmo. Nanofluid formulations based on two-dimensional nanoparticles, their performance, and potential application as water-based drilling fluids, *ACS Omega.*, 2022, **7**, 20457–20476.
[30] M. Sandhya, D. Ramasamy, K. Sudhakar, K. Kadirgama and W. S. W. Harun. Ultrasonication an intensifying tool for preparation of stable nanofluids and study the time influence on distinct properties of graphene nanofluids – A systematic overview, *Ultrason. Sonochem.*, 2021, **73**, 105479.
[31] R. N. Radkar, B. A. Bhanvase, D. P. Barai and S. H. Sonawane. Intensified convective heat transfer using ZnO nanofluids in heat exchanger with helical coiled geometry at constant wall temperature, *Mater. Sci. Energy Technol.*, 2019, **2**, 161–170.
[32] S. S. K. Kiu, S. Yusup, C. V. Soon, T. Arpin and S. Samion. Lubricant enhancement via hydrodynamic and acoustic cavitation, *Procedia Eng.*, 2016, **148**, 57–63.
[33] J. Joy, A. Krishnamoorthy, A. Tanna, V. Kamathe, R. Nagar and S. Srinivasan, Recent Developments on the Synthesis of Nanocomposite Materials via Ball Milling Approach for Energy Storage Applications, *Appl. Sci.*, 2022, **12**, 9312.
[34] F. Yu, Y. Chen, X. Liang, J. Xu, C. Lee, Q. Liang, P. Tao and T. Deng. Dispersion stability of thermal nanofluids, *Prog. Nat. Sci. Mater. Int.*, 2017, **27**, 531–542.
[35] R. Saidur, K. Y. Leong and H. A. Mohammed. A review on applications and challenges of nanofluids, *Renew. Sustain. Energy Rev.*, 2011, **15**, 1646–1668.
[36] S. Mukherjee, S. Wciślik, V. Khadanga and P. C. Mishra. Influence of nanofluids on the thermal performance and entropy generation of varied geometry microchannel heat sink, *Case Stud. Therm. Eng.*, 2023, **49**, 103241.
[37] M. Hemmat Esfe, S. Ali Eftekhari, A. Alizadeh, S. Aminian, M. Hekmatifar and D. Toghraie. A well-trained artificial neural network for predicting the optimum conditions of MWCNT–ZnO (10:90)/ SAE 40 nano-lubricant at different shear rates, temperatures, and concentration of nanoparticles, *Arab. J. Chem.*, 2023, **16**, 104508.
[38] M. Hemmat Esfe, M. Bahiraei and A. Mir. Application of conventional and hybrid nanofluids in different machining processes: A critical review, *Adv. Colloid Interface Sci.*, 2020, **282**, 102199.
[39] G. Ulrich. Flame synthesis of fine particles, *Chem. Eng. News.*, 1984, **62**, 22–29.
[40] A. J. Gröhn, B. Buesser, J. K. Jokiniemi and S. E. Pratsinis. Design of turbulent flame aerosol reactors by mixing-limited fluid dynamics, *Ind. Eng. Chem. Res.*, 2011, **50**, 3159–3168.
[41] K. Hembram, D. Sivaprakasam, T. N. Rao and K. Wegner. Large-scale manufacture of ZnO nanorods by flame spray pyrolysis, *J. Nanoparticle Res.*, 2013, **15**, 1461.
[42] W. J. Stark, A. Baiker and S. E. Pratsinis. Nanoparticle opportunities, *Part. Part. Syst. Charact.*, 2002, **19**, 306–311.

Index

https://doi.org/10.1515/9783111316819-006

www.ingramcontent.com/pod-product-compliance
Lightning Source LLC
Chambersburg PA
CBHW081529220326
41598CB00036B/6379